A LAB OF ONE'S OWN

PRAISE FOR *A LAB OF ONE'S OWN*

'an urgent and absorbing tale. Fara's impassioned yet rigorous work never falters or compromises in its search for a history that is both true and continues to matter a very great deal.'

Charlotte Sleigh, Professor of Science Humanities,
University of Kent

'Fascinating …[Patricia Fara] has uncovered the hidden, suppressed histories of scientists and clinicians who made great contributions to war and welfare, and she has woven a broader narrative of gain and loss that still resonates today.'

Jeremy Sanders, Former Pro-Vice-Chancellor
and Professor of Chemistry, University of Cambridge

'The stories in this book made me very happy that I came of age in the middle of the 20th century, when the world of science welcomed a woman's questions and valued her experiments.'

Maxine F. Singer, President Emeritus,
Carnegie Institution for Science

'A book full of fascinating insight and anecdote about women working in or with science around the time of the 1st World War. So many hidden stories and amazing heroines.'

Athene Donald, Professor of Experimental Physics
at University of Cambridge and Master of Churchill College

'Vividly and movingly, *A Lab Of One's Own*, brings to life the forgotten story of the scientific, mathematical, medical and technological contributions made by British women during the First World War, with legacies and lessons that still matter today. Patricia Fara deserves a medal.'

Gregory Radick, Professor of History
and Philosophy of Science, University of Leeds

PATRICIA FARA

A LAB OF ONE'S OWN

Science and Suffrage
in the First World War

OXFORD
UNIVERSITY PRESS

OXFORD
UNIVERSITY PRESS

Great Clarendon Street, Oxford, OX2 6DP,
United Kingdom

Oxford University Press is a department of the University of Oxford.
It furthers the University's objective of excellence in research, scholarship,
and education by publishing worldwide. Oxford is a registered trade mark of
Oxford University Press in the UK and in certain other countries

First Edition published in 2018
Impression: 1

Published in the United States of America by Oxford University Press
198 Madison Avenue, New York, NY 10016, United States of America

British Library Cataloguing in Publication Data
Data available

Library of Congress Control Number: 2017946189

ISBN 978-0-19-879498-1

Printed in Great Britain by
Clays Ltd, St Ives plc

To the memory of Delia Graff Fara

ACKNOWLEDGEMENTS

I slid into this topic almost accidentally, and I owe an enormous amount to the expert scholars whose work has guided me. In particular, I am very grateful to David Edgerton for encouraging me to move beyond standard views of British wartime science and for commenting on a draft chapter, and to Marsha Richmond, not only for her inspiring articles but also for giving me permission to appropriate one of her titles: *A Lab of One's Own*. In addition, I have greatly benefited from the helpful advice provided by anonymous reviewers.

To my surprise and delight, while I was writing this book, I was contacted by Viscount Davidson, whose grandfather played a crucial role in developing British wartime X-ray equipment. He has been exceptionally generous in sending me copies of family letters and photographs, and also for alerting me to the remarkable book-length memoir by Helena Gleichen, which I have never seen referred to elsewhere.

In May 2014, I presented a nascent version of this book at an international conference held at the Royal Society and organized by the Women in Science Research Network (WISRnet). I have greatly valued the support and enthusiasm of WISRnet's members, especially Claire Jones and Sue Hawkins, who subsequently edited a special issue of *Notes and Records of the Royal Society* (March 2015). Two other women were unwittingly important influences while I was writing this book: Karolyn Shindler, exemplary biographer of Dorothea Bate, and Elizabeth Crawford, whose marvellous posts on suffrage campaigners kept arriving in my inbox.

My agent Tracy Bohan, and my OUP editors Matthew Cotton and Luciana O'Flaherty, all showed extraordinary patience while I was completing the manuscript, and this book would certainly never have been published without their help. I am also very grateful to my meticulous copy editor, Henry MacKeith: I take full responsibility for any surviving mistakes.

I can only apologize to the many other friends and colleagues who have been extremely cooperative but will remain unmentioned in the interests both of brevity and of avoiding a confessional tale about my life during the last few years. There is, however, one exception—Clive Wilmer, who generously wielded his exacting red pencil on an entire draft, and tolerated many vacillations along the way.

CONTENTS

LIST OF ILLUSTRATIONS

LIST OF ABBREVIATIONS

CCC   Churchill College Cambridge

IWM   Imperial War Museum Women's Work Committee

LSE   Women's Library, London School of Economics

NCC   Newnham College Cambridge

NCC-WW   'War Work, 1914–18' compiled by Edith Margaret
             Sharpley and transcribed by Laura Archer-Hind, 1922
             (at Newnham College Cambridge)

Her full nature, like that river of which Cyrus broke the strength, spent itself in channels which had no great name on the earth. But the effect of her being on those around her was incalculably diffusive: for the growing good of the world is partly dependent on unhistoric acts; and that things are not so ill with you and me as they might have been, is half owing to the number who lived faithfully a hidden life, and rest in unvisited tombs.

George Eliot, *Middlemarch*, 1871–2

The past is never dead. It's not even past.

William Faulkner, *Requiem for a Nun*, 1951

He picked up the lemons that Fate had sent him and started a lemonade-stand.

Elbert Hubbard, *The King of Jesters*, 1915

PART I

PRESERVING THE PAST, FACING THE FUTURE

1

SNAPSHOTS

Suffrage and Science at Cambridge

'Gentlemen of the House of Commons.' This ancient phrase...is now out of date. Its superannuation is but one indication of the tremendous breach in Parliamentary tradition caused by the election to the Houses of Commons of Viscountess Astor....In the first place, where is Lady Astor to sit?...And, if she wears a hat, should she remove it when she rises to speak, as male M.P.s are bound to do?

The Times, 29 November 1919

In the summer of 1908, eleven young women with similar hairstyles posed for a photograph (Figure 1.1). Their demure demeanour implies that they were quiet and well-spoken, but the cricket bat and their striped ties reveal that they were far from conventional. For one thing, they belonged to an extremely small and select group: a century ago, there were twenty-three colleges at Cambridge, but only two of them accepted women. And cricket was a man's game.

These women appear well-scrubbed and modest, but their reputation had spread and their families were worried. Some of them may well have defied their parents' wishes by insisting on higher education; certainly all of them were unusually intelligent and resilient. One Newnham father complained that his previously sensible daughter had turned suffragist, reporting that 'there is a regular manufactory of very advanced women going on at Cambridge'.[1] He may perhaps have heard that a leading fomenter of suffrage unrest among the undergraduates was Newnham's cricket captain Rachel (Ray for short) Costelloe, who studied mathematics and here sits in the centre of

Figure 1.1. Cricket team, Newnham College, Cambridge, 1908.

her team holding the ball. Writing to an academic aunt in America, she enthused that 'we have started a society which has now got more than ¾ of the college, & which has joined with Girton & amalgamated to the national Society'.[2] Writing to former Newnham students, she collected 'cheques & postal orders & letters of sympathy.... Cambridge has become a centre of activity—meetings, debates, plays, petitions etc all through the term.'[3]

At first glance Costelloe appears indistinguishable from her team-mates, but she alone is wearing white shoes, not black ones. In another Newnham photograph, around fifty undergraduates pose in fresh white blouses—apart, that is, from Costelloe, who is the only one wearing a dark jacket.[4] Was this simply indifference to dress codes, or an urge not to conform, a decisive individualism that would characterize her future career as Mrs Ray Strachey, one of Britain's most prominent suffrage campaigners? Presumably not many of the other students in her cricket team spent July touring Scotland in a caravan 'to preach Suffrage'.[5]

Ensconced in the leafy grounds of Newnham College, these female athletes had little inkling either that the country would soon be engulfed in war or that only a decade later, women over thirty would be allowed to vote for the first time. Thanks to Costelloe/ Strachey and her colleagues, some of the women in this picture would be able to participate in the election that was held on 14 December 1918—provided, that is, that they had managed to survive the First World War. For over fifty years, campaigners (including some male supporters) had been demanding female enfranchisement, but their efforts intensified in the early twentieth century. As suffrage numbers swelled, various splinter groups emerged. Most suffragists favoured peaceful demonstrations—marches, lobbying, committee meetings— but militant activists engaged in increasingly violent protests. Derisively dubbed 'suffragettes', they came to dominate the headlines and the history books.

These privileged young cricketers had probably been educated in one of the newly fashionable establishments for girls. Modelled on the traditional boys' public schools, they catered for the demand from fathers who were worried about their family's financial future and regarded education for their daughters as a sound investment. For these schoolgirls, their well-educated teachers provided living evidence of a future different from following their mothers into domesticity. The possibility of financial and emotional independence from a domineering husband stood before them in classrooms every day.[6] Even so, going out to work could be still more frustrating than being trapped at home. Before Ida Mann managed to escape by training as a doctor, her father had sent her to Clarke's Business College. The routine clerical work was so boring that she used to lock herself in the lavatory, where she silently howled and bit her wrists. 'I once saw a young woman go mad at her desk,' she wrote; 'she was removed and never seen again.'[7]

As well as being taught to develop their intellectual abilities, girls were encouraged to take part in team sports to help them become the strong virtuous mothers that Britain needed for rearing leaders of the Empire. When Costelloe played in the school hockey team, she wore 'a red flannel blouse, and stiff white linen collar, a red and white necktie, and a straw sailor hat with a red and white ribbon band'.[8] Before going to Cambridge, she had already learnt from her best friend Ellie Rendel that 'The hockey here is very exciting the only thing wrong with it is that it's too engrossing. They have team practices continually and the game is so hard and fast that I can hardly move after one.... The great excitement at present is the freshers match.'[9]

In contrast, life ahead seemed less enticing. For the great majority of female graduates, the only realistic alternative to marriage and domesticity was teaching, and by 1911, there were already 180,000 unmarried teachers, often looked down upon as frowzy spinsters. The numbers increased after the War, when there were significantly more women of marriageable age than men. After a visit to Cambridge in 1928, Virginia Woolf wrote condescendingly about the students she had met: 'Intelligent eager, poor; & destined to become schoolmistresses in shoals', she confided to her diary. 'I blandly told them to drink wine & have a room of their own.'[10]

What women looked like mattered even more then than it does now. Long skirts, long hair, long sleeves—these restrictions confirmed that they belonged at home, looking after their men, their home, and their children. Yet although none of these sportswomen has yet dared to cut her hair, they have transgressed social norms by adopting a uniform. However unsuitably dressed they might seem for a strenuous game, their identical outfits advertise that they are a group apart. Over the next few years, many suffragists deliberately and provocatively chose to wear quasi-masculine clothes, knowing that most other people (women as well as men) disapproved of such unfeminine attire. Hostility towards women in uniform intensified during the Great War, when countless women took over men's jobs in factories and laboratories, in hospitals and the armed forces, in museums and transport.

By tradition, men routinely wore uniforms, both as protection and to signify their status. But whereas their official clothes earned them respect for their patriotic devotion to duty, women in khaki were mocked for abandoning femininity.[11]

These student cricketers appear innocent and old-fashioned, but they were emerging from a past that was already crumbling and they would help to create a new future. Cambridge men—students and dons alike—had been horrified when women arrived, first at Girton in 1869, and then two years afterwards at Newnham. Even in the early 1970s, there were still no colleges at either Cambridge or Oxford that accepted both male and female students. Of all the subjects studied at school, mathematics was perceived as being particularly unsuitable for women. In 1908, Dr Janet Campbell warned the Board of Education that schoolgirls should not be forced into this topic 'requiring much concentration and therefore using up a great deal of brain energy'. Instead, she recommended introducing cooking or embroidery, to reduce 'mental strain'.[12]

As a mathematics undergraduate, Costelloe symbolized everything that reactionaries found ridiculous about women's invasion of this exclusive male stronghold. *Punch* made sure that the entire country was alerted to the danger:

> The Woman of the Future! She'll be deeply read, that's certain,
> With all the education gained at Newnham or at Girton;
> She'll puzzle men in Algebra with horrible quadratics,
> Dynamics and the mysteries of higher mathematics.[13]

Mockery is a favourite defence tactic of the frightened, and only six years after this verse appeared in 1884, male fears of being beaten in examinations were realized when Philippa Fawcett came top in the finals examinations for mathematics. A Girton student described travelling to central Cambridge one sunny morning to hear the results. Looking down from the packed gallery of women, she watched the men on the ground floor rowdily cheering the students who had done best—the Wranglers—and banging wooden spoons for the lowest.

And then, the long awaited moment arrived. The Moderator stood there waiting till the noise was hushed, and then the words rang loud through the Senate House: "Above the Senior Wrangler, Fawcett, Newnham." The shout and the applause that rang through the building was unparalleled in the history of the University.'[14]

To commemorate this momentous achievement, one of Fawcett's friends at Newnham wrote a five-stanza poem. Although its rhyming is as painful as the verse in *Punch*, its sincerity rings clear:

> Hail the triumph of the corset
> Hail the fair Philippa Fawcett
> Victress in the fray
> Crown her queen of Hydrostatics
> And the other Mathematics
> Wreathe her brow with bay.[15]

Despite this 'triumph of the corset', male honour was preserved. Women were not allowed to participate in the Cambridge graduation ceremonies until 1948, but instead were sent a certificate through the post to confirm that they had completed the course. Fawcett was never officially awarded her degree, and never received the traditional accolade of being named First Wrangler. Oxford managed to appear less hostile: there were four all-female colleges, the class lists were arranged alphabetically, and they allowed women to graduate formally in 1920. But after heated debates, Cambridge refused to shift. To celebrate their victory, male opponents to the change converted a handcart into a makeshift battering ram, smashing it into Newnham's bronze entrance gates and partly destroying them.[16]

Perhaps mindful of its gates, Newnham College carefully safeguards its paper records of the past. When I visited the library, the archivist proudly showed me a substantial hand-made book with embroidered linen covers that records the activities during World War One of around six hundred Newnham members (see Figure 1.2). Some of them were recent graduates, others were well into their fifties. Inscribed in exquisite red and black lettering on luxuriantly thick pages are the names of doctors who operated at the front, chemists who developed

Figure 1.2. Edith Margaret Sharpley and Laura Archer-Hind, 'War Work, 1914–18'. Manuscript at Newnham College, Cambridge.

explosives and poison gases, biologists who researched into tropical diseases, and mathematicians recruited for intelligence work. Some of them died on service abroad; and many were rewarded with government or military honours, not only from Britain but also from Serbia, France, Russia, Belgium, and Romania. Costelloe's friend Ellie Rendel appears here not as an ardent hockey player, but as a medical worker at the Eastern Front who was decorated with the Serbian Order of St. Saba and the Russian Order of St. George and St. Anna.[17]

This unique volume, lovingly created and preserved, offers a rare and poignant glimpse of the vital contributions made by scientific women all over Europe during the First World War. From Newnham alone, there were women investigating industrial fatigue, carrying out ballistics calculations, inspecting factories, investigating how vitamins might be preserved in sun-dried vegetables, testing radio valves, improving high-quality glass, making artificial limbs, compiling statistics on sugar production, and testing steel. The very first page includes a physicist who ran hospital X-ray departments, a mathematician who travelled to Serbia as a doctor, and a scientist who survived a typhus epidemic abroad but died of pneumonia in London soon after returning home. If similar records had been kept at other universities, they too would have revealed the wartime work of many other remarkable women.

Gingerly leafing through the pages, I wondered why I recognized none of the names—and after carrying out more research, that question continued to haunt me. Why had I heard of Vera Brittain and Edith Cavell, but not the chemists who were experimenting with explosives and mustard gas in trenches at Imperial College? Why does our country not commemorate the female doctors who ran hospitals under atrocious conditions in Salonika, or the London University professor of botany who headed the women's army in France? Why are these extraordinary women absent from the numerous books detailing the scientific, medical, and technological advances spurred by the War? I resolved to create my own tribute to these scientific pioneers who had helped win the War and gain the vote.

Counter-intuitively, the country's leading militant suffragette, Emmeline Pankhurst, greatly admired Thomas Carlyle, the Victorian thinker who declared that the history of the world is the history of great men. His way of thinking dominated attitudes for many decades, but fashions changed. During the second half of the twentieth century, Carlyle's version of the past lost popularity with historians, who now preferred a 'bottom-up' approach. They began crediting the lives and achievements of ordinary people—first men and then women. As a consequence, there are now many marvellous books about the female ambulance drivers, munitionettes, nurses, bus conductors, farmers, factory workers, and all the other women who effectively ran Britain for four years while their menfolk were away.

A Lab of One's Own contains some more unusual and less familiar stories; ones about highly trained female scientists and doctors. Although relatively few in number, these women had a huge impact. They tested insecticides, researched into tropical diseases, analysed codes, invented drugs, ran hospitals...yet despite all these contributions, their existence and their activities have been eclipsed. These scientific pioneers have suffered from a double neglect in the nation's memory. First, they are largely ignored in studies of women during wartime: feminist historians have understandably preferred to focus on wartime manual workers, who were far more numerous and left behind more readily accessible evidence. At the same time, professional women are scarcely mentioned in books about science and medicine during the War, which deal almost exclusively with men.

By 1914, women had been studying science at university for almost half a century, often braving opposition and mockery from fellow students as well as from parents and teachers. The ones who persevered had already demonstrated that they were exceptionally intelligent, courageous, persistent, and full of initiative. At the beginning of the War, many of them immediately started looking for opportunities to serve their country, even if that entailed sacrificing their own careers in the process.

As men left for the front, first voluntarily and then under conscription, scientifically trained women took over their vacated positions in museums, boys' schools, and government departments. Those already engaged in research abandoned their current projects and switched their attention to the essentials of warfare, such as explosives, drugs, insecticides, alloys, glass, and aircraft design. Others emerged from back-room positions, appearing for the first time on the lecturer's podium or at the experimenter's bench. Medical schools temporarily welcomed female students, and some doctors defied government recalcitrance by going to serve overseas, where they endured exceptionally demanding circumstances. Benefiting from the unusual experience of having worked professionally in one of the few environments where men and women could be equally well qualified, if not equally well recognized, some female scientists attained high administrative positions. For four years, these scientists, doctors, and engineers proved not only that they could take over the work of men, but also that they could often do it better.

'The war revolutionised the industrial position of women,' proclaimed the suffrage leader Millicent Fawcett; 'It found them serfs, and left them free.'[18] But many historians no longer celebrate the War as a dramatic turning point for women, and instead interpret it as a temporary opening up of opportunities before the door clanged shut once again. At the time, feminist opinion was divided. Like Fawcett, some former suffragists were jubilant, regarding 1918 as the dawn of a new era when women had at last won the vote and could work side by side with men in conventionally male occupations. But others felt that although women had helped to win the War, they had lost the battle for equality. Conventional hierarchies were rapidly re-established, prompting campaigners to protest that the vote was of little practical use in the absence of professional parity.

As unemployment levels rose during the 1920s, men were given priority in the scramble for jobs. Women who had run hospitals or supervised research teams were forced back into the same low-status positions as before the War, and many factory workers were laid off. Once they had experienced the relatively high wages and freedom of munitions work or farming, women were reluctant to resume their earlier lives in domestic service. Former servants either refused to take back their old posts unless they were offered higher wages and regular time off-duty, or else migrated into shops, offices, and the new light industries.

Under these circumstances, having the vote made little practical difference—but women themselves had changed. They now knew that collectively they were capable of undertaking the same responsibilities as men, even if they were prevented from doing so for the present. As Vera Brittain put it, although 1919 was a terrible year, it was also 'the spring of life after the winter of death...the gateway to an infinite future—a future not without its dreads and discomforts, but one in whose promise we had to believe, since it was all that some of us had left to believe in'.[19] Many people believed that the route to that golden prospect lay through science.

What happened a century ago is important for understanding the present. In science, as in other fields, men returning from the front reclaimed jobs that women had been competently carrying out in their absence. Yet at the same time, the government poured money into science, industry, and education after the War, so that although the old stereotypes and prejudices re-emerged, there were more possibilities for scientific women than ever before. They were still not getting their fair share, but at least the cake was larger. Even if the War did not completely overturn old hierarchies, women had proved themselves capable of carrying out work traditionally reserved for men, and—try as reactionaries might—that unprecedented sense of achievement and power and possibility could not be obliterated.

British society had been indelibly altered, so that although the future seemed unsettled, there was no going back to the fixed and oppressive certainties of the past. My mother was born in 1918—the year women aged over thirty gained the right to vote—and she inevitably passed on to me the hopes as well as the fears of her generation. Like many of my own contemporaries, I have tried to continue the pioneering initiatives of the women in this book and to improve the position of scientific women. Old prejudices still resonate through society, and glass ceilings still restrict women's rise to the top, but the extent of change has been enormous, even if the pace remains frustratingly slow.

There are as many ways to tell a story about World War One as there were people in it. Each would be in some sense right, although none would give a complete picture. A *Lab of One's Own* does not pretend to provide a definitive version of World War One. But what it can do is offer new ways of thinking about the early twentieth century by looking simultaneously at the involvement of science and of women. Often only snippets of information about individuals survive, but I hope this book can act as a paean to the pioneering scientists and doctors whose lives are now obscure but who collectively made my own professional career possible.

2

A DIVIDED NATION

Class, Gender, and Science in Early
Twentieth-Century Britain

Go home, attend to your own handiwork
At loom and spindle
As for the war, that is for men.

Homer, *The Iliad*

America and England—two nations divided by a common language.
Opinions vary about whether it was George Bernard Shaw,
Oscar Wilde, or even Winston Churchill who said that first, but
whoever it was, he could with equal justification have said that Britain
was itself a divided country. In that supposed democracy, before the
First World War only a third of the adult population had the right to
vote. Around half of Britain's inhabitants were automatically disqualified because they were women, but even men needed to be property
owners aged over 21. Like women, the poor remained unrepresented.

Money, class, and gender all mattered. In a single telephone call, one
wealthy woman from Kensington ran up a grocery bill of £65—a year's
income for a family living only a few miles away in Lambeth.[1] A genteel
campaigner for the vote shuddered on encountering lower-class realities. 'Breakfasts in this sort of abode are impossible', she complained
when obliged to stay in cheap London digs; the maid 'makes me feel ill
every time she comes near. So filthy, so mean, so little like a human
being...but, oh, her presence and her hands!!! Ye Gods!!!'[2]

Women were valued less than men and they were paid less, even for
the same job. A Civil Service report of 1913 spelled out the biological

14

justification: 'Men command higher salaries than women because they are worth more, being stronger and capable of getting through more work in a day. Probably their judgement is sounder; at any rate they are preferable generally for positions of power.'[3] Frequently repeated, such pronouncements were internalized as scientific facts. Emmeline Pankhurst was once surprised to be told that she was sitting next to a female engineer at a formal dinner. Her tone of shocked refinement is almost audible: 'But surely, that's a very unsuitable occupation for a lady, isn't it?' she exclaimed. Her use of the word 'lady' suggests that even this eminent suffragette had failed to cast off ingrained assumptions about class and gender.[4]

There were also great geographical disparities. In 1913, university physicists were using the latest high-tech equipment to study atoms and X-rays—yet at the same time, the remaining tenant farmers in the Outer Hebrides were scraping an existence by bartering woollen cloth that they had woven by hand beneath the murky light cast from lamps burning seabird oil.[5] Similarly, while Virginia Woolf and her middle-class London friends fretted about what clothes to wear for their next dinner party, in Lancashire female cotton weavers not only held down full-time jobs but *also* looked after their houses and families:

> On Monday I clear up all the rooms after Sunday, brush and put away all Sunday clothes, and then separate and put out to soak all soiled clothes for washing. On Tuesday, the washing is done.... Saturday morning is left for all outside cleaning—windows and stonework—besides putting all the clean linen on the beds.[6]

A few stark statistics reveal the realities of a nation that was even more unequal than it is now. The richest 1 per cent of the population controlled 70 per cent of the national wealth. Eight million people lived on less than £130 a year, although the Chancellor of the Exchequer's annual salary was £5,000. The average age of death was fifty-five in London's fashionable West End, but thirty in the poorer East End. Only 6 per cent of sixteen year olds were at school (after the War, the school-leaving age was raised from 12 to 14). In 1900, there were nearly 2,000,000 women in domestic service, but only 200 female doctors and two architects.[7]

Facing a Common Enemy

In the summer of 1914, the country was riven by internal battles about industrial exploitation and governmental inequities. Repeated strikes were threatening national productivity—even schoolchildren were marching in protest against their teachers. By the autumn, the Woolwich Arsenal would be taking on extra staff to provide explosives needed for international warfare, but in July, it was closed for several days during a violent dispute about union membership. Adding to this industrial unrest, the Irish were pressing for Home Rule, while militant suffragettes were smashing shop windows, slashing pictures, and assaulting police; in return, they were themselves being attacked physically as well as verbally.[8]

In some ways, August's declaration of war relieved these explosive tensions. Seizing the opportunity to silence perceived troublemakers—striking workers, Irish rebels, militant suffragettes—leaders encouraged loyal Britons to unite against the common enemy. Never one to miss the opportunity for a memorable turn of phrase, Millicent Fawcett promptly sent out a stirring message to the National Union of Women's Suffrage Societies (NUWSS). 'Women, your country needs you', she declared, two days before Lord Kitchener's renowned recruiting slogan 'Your King and Country Need You' first appeared in the press.[9] Writing in *Nature*, Nobel Prize winner William Ramsay set out to rally British scientists and advertise the importance of their research for the national cause. 'The doctrines of humanity are deeply rooted in the Anglo-Saxon spirit ... this is a war of humanity against inhumanity', he urged; 'the restriction of the Teutons will relieve the world from a deluge of mediocrity.'[10] 'However the world pretends to divide itself', wrote Rudyard Kipling in the *Morning Post*, 'there are only two divisions in the world today—human beings and Germans.'[11]

To an extent, the demands of warfare temporarily blurred differences of class, if not of gender. Almost a million British soldiers died, and they included not only uneducated farm workers, servants, and factory hands, but also privileged volunteers, such as Britain's most famous war poet—Wilfred Owen—and Henry Moseley, a young Oxford physicist predicted to win the Nobel Prize for his X-ray research. Women from different backgrounds also came together as never before. A high-society journal reported that 'the wife of an officer, coming home from Egypt, decided to volunteer'. More extraordinary still, 'her maid, who had travelled with her, followed her into the works, and now the two go to work of a morning together and return together at night'.[12] Standing at a factory bench, a Girton graduate was flanked by 'a soldier's wife from a city tenement [and] a vigorous daughter of the Empire from a lonely Rhodesian farm'.[13]

In laboratories and hospitals, all-female teams collaborated to replace the men who had gone to fight, and by the end of the War almost a million women were carrying out chemical work in munitions factories across the country. Similarly to soldiers, they were united by their repetitive work schedule and their uniform dress— not khaki, but blue overalls with mob caps to protect their hair. Rapidly dubbed 'munitionettes', many of them had previously worked in northern textile mills, but they were also drawn from remote Irish villages and the basements of grand country houses. As employers remarked, some of them also came from the drawing rooms of suburban villas:

> in one large Engineering Works in Reading I know that a great deal of Shell Making has been done by Women, several of them of gentle birth, and that an arrangement has been made amongst these Ladies to take turn about so as to do between them whole time night and day for one two or more workers.[14]

An overseer at the Woolwich Arsenal reported that her shift included 'three fever nurses, two dressmakers...two cooks, a lady's maid, two sisters who kept a boarding house with their mother...several clerks [and] two or three married women with no children'.[15] Embarrassed by their idleness in a time of emergency, privileged ladies became 'sick of frivolling, we wanted to do something big and hard, because of our boys, and of England'.[16] Having signed up for tedious routine work, one well-off employee 'felt a curious satisfaction and happiness in being just an ordinary worker pushing my way through the gates with hundreds of others'.[17]

Even so, snobbishness did not evaporate. One woman who delighted in witnessing duchesses and coster-girls crammed together on a production line, all dressed in the same earth-coloured overalls, retained her ingrained tones of condescension. 'The ordinary factory hands have little to help keep them awake.' she reported. 'They lack interest in their work because of the undeveloped state of their imaginations.'[18] Ethel Brilliana Tweedie, an eloquent campaigner for female education and an early Fellow of the Royal Geographical Society, had nothing but praise for wealthy women who had thrown themselves into war work, but complained about the uppity servants who demanded their wages. According to her, 'Women with brains and education cannot organize and run charities, canteens, or hospitals if the cook neglects their dinner', and she griped about the selfishness of under-paid, over-worked servants who had deserted in droves for factories and offices; 'A nice little dainty-looking dinner served up to her brain-fagged mistress, is the cook's bit of war work, and is a real help to the country.'[19] After all, she pointed out, for the recently impoverished rich, 'the enormous responsibility of an expensive house, several servants, and children to be schooled, hang like leaden weights round the necks of super-taxed families'. Seemingly oblivious to the realities of working-class life, she explained that:

> One has only to deal with the lower orders to understand why they are poor. They waste more good material than rich folk utilise, and then they grumble at poverty. They throw away vegetable water, fish water, stewed meat water, and seldom or never realise that one and all are the groundwork of good nourishing soup.[20]

For four years, the cracks in the divided nation were temporarily papered over, but gross inequalities of class and gender soon became visible again. For the dwindling population of isolated crofters in the Outer Hebrides, military service had offered a welcome escape route from a life of hardship, but it was not until decades later that tourism and government investment revived the local economy. In contrast, living in fashionable Bloomsbury and newly entitled to vote, Virginia Woolf continued the literary career she had started in 1900, which included writing for *Vogue*. Although rejoicing in her independence, she also grumbled about the increased difficulty of getting the servants she needed to maintain it.

Women and Difference

Women are still often discussed as if they were a homogeneous minority. But just like men, they make up half the population and—unsurprisingly—did not and do not share the same opinions. Britain's first female Member of Parliament, Lady Astor, reassured the House that there would never be a unified female party: 'We could not do it. We women disagree just as much as the men.'[21] One set of disputes revolved around the question of whether the two sexes (the term 'gender' became fashionable only in the 1970s) were intrinsically different from each other. Another fundamental split concerned opportunities. Should women participate equally with men in a shared working environment, earning the same amount and entitled to the same privileges? Or should women be financially rewarded for staying at home to bring up their children? Should male and female students be taught together, or would female students flourish under a separate regime free from male oppression?

Those women were not fighting for the same freedoms as modern feminists, and even among suffrage campaigners there was no uni-fied view. Consider Millicent Fawcett, daughter of a coal merchant, sister of the pioneering doctor Elizabeth Garrett Anderson, and mother of Philippa, the undergraduate who vindicated the value of female education by coming top in Cambridge's mathematics exam-inations. Her credentials seem impeccable: she helped to found

Newnham College and was a key player in launching the suffrage movements of the late nineteenth century. But during the War, she supported a view that now seems horribly reactionary, arguing that while 'the necessary, inevitable work of men as combatants is to spread death, destruction... sorrow untold, the work of women is the exact opposite... to help, to assuage, to preserve, to build up the desolate home, to bind the broken lives, to serve the state rather than destroying it.'[22]

Many people—women as well as men—agreed with Fawcett that differences between the sexes were historically indisputable and scientifically established. When the War started, it seemed self-evident that women should be kept well away from any military action. While soldiers went off to defend their country overseas, women were to remain behind, tending that domestic fortress known as the Home Front. 'Every fit woman can release a fit man' declared army propaganda, thereby implying that they were required as second-class replacements for more valuable human beings. Some suffragist slogans resemble government advertisements. A recruitment poster for Queen Mary's Auxiliary Army Corps featured 'The GIRL behind the man behind the gun'—but in an earlier march organized by the Pankhursts, many of the 20,000 protesters had carried banners voicing exactly the same sentiment: 'Shells made by a wife / May save a husband's life.'[23]

A Room of One's Own (1929) is now regarded as the seminal feminist text of the early twentieth century. In contrast, when Virginia Woolf had visited Cambridge University the previous year to present an early version, her female audience was unimpressed by this novelist in her forties. One student confessed that she slept right through the lecture, and another was disappointed not to hear 'some profound, philosophical remarks, even after prunes and custard'. Woolf herself was relieved to get home. 'I felt elderly & mature. And nobody respected me', she confided to her diary.[24]

Woolf insisted that there is an essential difference between male and female creativity. Although rarely discussed, there is a scientific interlude in *A Room of One's Own* that was probably inspired by her friend Janet Vaughan, a physiologist whose research entailed mincing up raw liver to investigate pernicious anaemia. Woolf described an imaginary novel about two young women working together in a medical laboratory. The fictional narrator 'read how Chloe watched Olivia put a jar on a shelf and say how it was time to go home to her children. That is a sight that has never been seen since the world began, I exclaimed.'[25] One easy interpretation is that Woolf was envisaging a utopian future in which the overwhelmingly male domain of science had been successfully penetrated by women. Nothing is straightforward in Woolf's writing, but the context suggests that she was more interested in exploring how writers might represent women in their own world, with their own motives and interests, rather than defining them solely in relation to men.

Many real-life female scientists opposed Woolf's ideal of separate worlds. Their ambition was to stand alongside—or even over—men and to work on the same projects. After the head of the Department of Botany at Birkbeck College, Helen Gwynne-Vaughan, was recruited for military work in 1917, she commented that female scientists were unusually well-qualified because they had experience of 'perhaps the only sphere in which at that time young men and women worked freely together—the laboratories of a modern university'.[26] This might well have been wishful thinking, but it did represent the idealized situation envisioned by progressive scientists. In one of his novels, H. G. Wells painted a similarly sympathetic picture of the laboratory as a secluded space, an egalitarian zone with its own conventions where female encumbrances were left behind at the door to the outside world. 'We have no airs and graces here,' said Wells's heroine as she clicked shut her microscope case; 'my hat hangs from a peg in the passage.'[27]

But whatever their aspirations, women training to be scientists went through very different experiences from men. Until the 1970s, at Oxford and Cambridge female students were segregated in their own colleges—and before the First World War, they were often also excluded from mixed lectures, practical classes, or field trips. In response, Newnham had set up its own single-sex Balfour Laboratory for the life sciences, which ran for thirty years before being closed down in 1914. It produced many of the twentieth century's most eminent female scientists, although they were not able to graduate fully until 1948.[28]

Female students fared far better at the relatively recent red-brick universities such as Birmingham, Bristol, Leeds, Liverpool, Manchester, and Sheffield. After the child psychologist Susan Isaacs left Manchester for Newnham in 1913, she remarked with incredulity that she had entered a seminary in which a chaperone had to be present even when her brothers came to visit. As an anonymous female student put it, women entered the new universities in their formative stages, whereas at Oxbridge they were intruders in ancient institutions. 'It is not good for the undergraduate to feel that women are present in *their* University only on sufferance,' she wrote; 'Nor is it good for the women students to be constantly reminded of the fact that they are not merely students but also female ones.'[29]

Before the War, around a quarter to a third of the students in the newer civic institutions were women, generally (but certainly not exclusively) drawn from lower down the social spectrum than at Oxbridge.[30] Degrees had been fully open to women well before the end of the nineteenth century, and because these universities were designed to serve local communities, many students lived at home rather than in colleges or halls of residence. As a result, the university was relieved of parental responsibilities, and women were much freer to do as they wished. The comparative lack of public scrutiny gave female students at red-brick universities more opportunities than at Oxbridge to run their own lives. The press was uninterested in the activities of relatively unprivileged women in provincial cities, so that despite being relegated to separate common rooms, they could mingle with men in libraries, practical classes, and lecture rooms, and play prominent roles in student societies. The influence of these enter-prising and comparatively liberated students often exceeded their numbers. For example, at Bristol University in 1909 the women's students union clubbed together with the men's to finance the physics and the chemistry societies, and it was at Manchester that the British Federation of University Women was launched in 1907. Already accustomed to thinking for themselves, these forceful graduates were well-prepared to take over during the War.

The Impact of War

When she lectured to those disappointed Cambridge undergraduates, Woolf shared an optimistic view of social progress that was widespread in the late 1920s. A pacifist, she openly opposed the War while it was happening, but later came to hail it as a dramatic break with the past, a turning point that had benefited everybody by banishing forever the traditional hierarchies and stifling conventions of the Victorian and Edwardian eras. As Woolf perceived it, the War had both unshackled women politically and liberated them artistically. But not everybody endorsed these confident claims made by an affluent married woman. Ill for some of the War, she had been spared the extreme privations suffered by so many others, and her personal difficulties had revolved around rationing, air raids, and the difficulty of finding servants.[31] As seen from less privileged perspectives, the War did not permanently transform the position of women. Although they now knew from experience that they were capable of fulfilling the roles traditionally reserved for men, such inner pride was not much help in securing permanent employment and a pay packet—basic realities about which Woolf knew little.

When war was declared in August 1914, most people—including the nation's leaders—thought that it would be over within a few months. Patriotic suffragists were quick off the mark, immediately launching recruitment and training campaigns for women well before the government became involved. Initially, female employment plummeted as economic belts were tightened, but once it became clear that hostilities were going to last for some time, the shift towards industrial work accelerated. Women found themselves in demand to take over jobs at home left vacant by men who had volunteered (conscription was first implemented in March 1916). Suffragists prematurely rejoiced that 'woman's place, by universal consensus of opinion, is no longer the Home. It is the battlefield, the farm, the factory, the shop.'[32]

'To-day every man is a soldier, and every woman is a man.... [W]ar has turned the world upside down; and the upshot of the topsy-turvydom is that the world has discovered women, and women have found themselves.'[33] Like all memorable slogans, this paints an appealingly simplistic picture. The major business of the country was warfare, and that was seen as a job for men. Their wives, sisters, and daughters provided a cheap solution to the problems of manufacturing ammunition, nurturing the next generation and looking after the wounded. Women were welcomed not for their own sake, but to provide temporary, inferior, and cheaper replacements for men. When women took over positions that had previously been held exclusively by men, the technical term was not 'substitution' but 'dilution', on the assumption that the workforce was being weakened. Infuriated, an Oxford undergraduate in her early twenties—now famous as Vera Brittain—protested that 'women in war seemed to be at a discount except as the appendages of soldiers'.[34]

The majority of women entering paid employment were relatively uneducated. Many seized this opportunity to escape from domestic service, where they had been financially exploited and stripped of their independence, and instead take on traditionally male tasks such as driving buses, operating factory machinery, performing clerical duties, delivering coal, and running farms. Lacking equipment such as washing machines or vacuum cleaners, these working women still had heavy domestic responsibilities, so that many of them were engaged in hard physical labour both at work and at home. But although they had stepped up promptly to take over menial, repetitive, and dangerous work, they were mostly dismissed after the Armistice, regarded suspiciously as extravagant opportunists taking the rightful wages of fathers and husbands.

These women were, however, transformed by the War. While she was serving in the Land Army, Annie Edwards liberated herself not only physically but also symbolically. One hot summer day during harvesting, in a rite of passage towards freedom, she cast off her corsets, the concealed restrictions of her former life:

> [W]hen I put the horses in the stable at twelve o'clock I went up in the cake loft and I took [my stays] off. And close to the farm there was an outside lavatory, you know, where they was emptied every third year or something. They put it on the land for manure. Well, I went and folded them and took them there. And from that day to this I never wore anything at all.[35]

At the other end of the social scale, Vera Brittain acted like many other protected young women, adding three years to her age so that she could volunteer as a nurse, rejoicing that 'After twenty years of sheltered gentility...I was at last seeing life.'[36]

These women helped to convince the country that women could indeed be responsible citizens. As one journalist put it, their patriotism had demonstrated 'women's right to share in the future of a nation whose very fate is entwined with their very heart strings'.[37] Although chauvinist diehards insisted that because women had not suffered in the trenches they did not deserve to control the country, they were outnumbered. Viscount Peel told the House of Lords in 1917 that 'many have been converted by the services rendered by women during the war'.[38] For example, as Prime Minister before the War, Herbert Asquith had been adamantly opposed to female suffrage, but by March 1917 he was telling the House of Commons that 'there is hardly a service...in which women have not been at least as active and as efficient as men, and wherever we turn we see them doing...work which three years ago would have been regarded as falling exclusively within the province of men.'[39]

Like Woolf, many historians favour a model of definitive rupture. But others maintain that after the War, women lost all the gains they had made. They cite the returning soldiers who were scathing about the '[m]illions of girls' enjoying themselves 'in some kind of fancy dress with buttons and shoulder-straps, breeches and puttees'.[40] Consumed by nostalgia, these men yearned to restore a lost golden paradise that had never in fact existed. Anxious about their own financial security in a time of rising unemployment, they resumed their former hostility towards women who entered the workplace. Although some women could now vote, runs this type of argument, power at home still belonged to the man in the family.

But in reality, nothing could be quite the same again. To a greater or lesser extent, every single person in the country was affected by the War. Inevitably, because so many women had been so active, the War helped to change opinions about their abilities and their roles. Surprised recognition seeps out from this 1919 letter preserved in the archives of London's Imperial War Museum:

> Offices too, in which before the War employers would not dream of having girl clerks, have been almost entirely staffed by girls, and I believe that there [sic] work has been well done, and they have undertaken jobs which have been done by men, and in many instances done them quite satisfactorily.[41]

Although many people ostensibly resumed their pre-War lives, what it meant to be a man and what it meant to be a woman had altered for ever.

Commemorating the War

By gaining the vote, women became a new feature in the political landscape, and at last they began to be visible in national commemorations. Britain's official war artists had chosen to overlook a third of the country's employees, but in 1918 the Imperial War Museum commissioned Anna Airy to produce four pictures showing women engaged in heavy wartime labour—building aeroplanes, running gasworks, making munitions. For one enormous canvas—over two metres wide—she visited a converted Singer sewing-machine factory in Glasgow.

Figure 2.1. Anna Airy, *Shop for Machining 15-inch Shells: Singer Manufacturing Company, Clydebank, Glasgow* (1918).

Her painting, *Shop for Machining 15-inch Shells*, reveals the appalling conditions that had made pre-War Scottish industry so profitable but prompted strikes among the men working there. It portrays an all-female uniformed workforce handling shells 15 inches in diameter and several feet long. Having previously produced domestic equipment to keep women busy at home, this industrial plant now employed former housewives to manufacture instruments of death.

Novels and films often idealize the Great War by depicting it as the Great Watershed, and by featuring simplistic stereotypes of courageous men and nurturing women.[42] The reality was different. For one thing, social structures had been changing for many decades, and they continued to do so after the War. Rather than reversing entrenched attitudes overnight, the War made earlier shifts apparent and enabled change to continue—it revealed and accelerated processes of transformation that had begun previously. The women who developed explosives in laboratories or performed emergency surgery on the battlefield or intercepted wireless signals could only do so because their predecessors had already been hammering away at conventional barriers.

Whether male or female, war heroes resemble fictional characters rather than real-life people. These mythologized caricatures feature in romanticized biographies corroborating conventional gendered visions of behaviour. Whereas men are commemorated for their bravery, initiative, and leadership, iconic women display feminine virtues of caring and victimhood. The most famous memoir, Vera Brittain's *Testament of Youth* (1933), appeals to readers by portraying two stereotypes—the naïve upper-class volunteer who undertook menial tasks in French hospitals, and the lovelorn faithful innocent who lost her fiancé to the patriotic cause.

In contrast, working women were overlooked then, and still can be today. In 2014, a below-stairs exhibition at Alnwick Castle paid loving tribute to the male gardeners and grooms who had signed up to fight at the front, but not a single female servant was mentioned. It seems that uneducated factory workers who perform hard labour carry less sentimental appeal then nice young ladies who instinctively step into a caring role. The suffragist Mabel St Clair Stobart, who ran field hospitals in Belgium and Serbia, refused to be deceived, commenting that these volunteers were 'a placebo—a bread pill, to dupe women into the belief that they are being taken seriously by the War Office . . . a farce—a mere drawing-room game, conducted upon the principle that women are incapable of anything but amateur nursing'.[43]

Of all the many thousands of suffrage campaigners who marched for the vote, only one is a household name—Emmeline Pankhurst. Her militant tactics alienated the British nation, split the suffrage movement apart, and led to many injuries and even deaths. Yet reverence would not be too strong a word to describe public sentiment towards her. Could it be, perhaps, because she cannily insisted that her suffragettes look as feminine as possible? Not for her the cropped hair and uniforms that made other campaigners vulnerable to accusations of not being real women: she gained a reputation for gentleness and charm that completely belied her aggressive actions.[44] In contrast, the educated women who spurned Pankhurst's tactics and donned masculine uniforms to carry out men's jobs—surgery, chemical research, aeroplane design—are scarcely remembered. Winston Churchill paid lavish tribute to the pioneering Scottish doctor Elsie Inglis, who—ignoring War Office opposition—led teams of medical experts to set up hospitals and laboratories in Serbia. She and her female staff would, he declared, 'shine forever in history'—but the Edinburgh Hospital named in her honour no longer exists.[45]

3

SUBJECTS OF SCIENCE

Biological Justifications of Women's Status

'What beautiful flowers you have in the drawing-room,' she said.
'Nothing remarkable, my dear. Everybody has flowers in their drawing rooms—they are part of the furniture.'
'Did you arrange them yourself, aunt?'
'The florist's man,' she said, 'does all that. I sometimes dissect flowers, but I never trouble myself to arrange them. What would be the use of the man if I did?'

Wilkie Collins, *Heart and Science* (1883)

At the beginning of the twentieth century, science represented modernity and progress towards a better future—and that made it an ideal symbol for suffrage campaigners. On 14 June 1908, female graduates donned their academic robes to join over 10,000 women in a grand procession that gradually threaded its way through London to the Albert Hall. Marching with pride and dignity, these determined suffragists won the approval of the crowds lining their route. The small band of scholars attracted particular attention. Annie Kenney, the only working-class woman to reach the upper echelons of suffrage organizations, commented that 'graduates always, I noticed, awed the public. A woman in cap and gown roused great admiration.' Carrying banners to celebrate major pioneers such as Marie Curie and Elizabeth Blackwell, scientists and doctors demonstrated to the crowds lining the streets that women were fully capable of professional careers: they were visibly *not* the abnormal freaks caricatured in the press.[1]

Yet science could also be seen as an agent of oppression. A head-on conflict between science and citizenship flared up three years later, when suffragists protested against being recorded in the 1911 census and urged their members not to cooperate. Since the government refused to let women's voices count, they objected, women should not let themselves be counted as if they were mere pieces of data. Edith How-Martyn, a suffragist with a degree in physics and mathematics, told *Times* readers that 'the Census is designed, not by a scientist for scientists...but for politicians with the knack of juggling with statistics'. Campaigners encouraged women to elude the census takers by being absent from home when the forms were completed, instead taking shelter in special refuge centres with all-night festivities. But not everyone in favour of suffrage agreed that this was the right strategy. *Times* readers were also informed about the opposite argument, that 'to boycott the Census would be a crime against science'. According to this view, refusing to comply with the census takers would be counterproductive because the state needed statistical information in order to improve the position of women: they should cooperate with this scientific attempt to arrive at accurate knowledge.[2]

However bitter the sentiments, this was a short-lived struggle, but it reflected deeper antagonisms. Those very women who embraced scientific ideas were adversely affected by them. Science is often celebrated as the neutral voice of reason overturning old prejudices, yet despite the ideological claim that theories are formulated objectively on the basis of solid evidence, they can be used to justify the existing *status quo*. In the early twentieth century, science endorsed rather than reversed older assumptions about female inferiority.

Looking back over past centuries, it now seems easy to identify the loop of faulty logic that results in scholarly support for sexual determinism.

Step 1: Observe that in human society, men are judged to be superior to women.

Step 2: Incorporate that assumption within a scientific theory about the natural world.

Step 3: Insist that because the theory is scientific, male superiority must be inherent in nature.

Step 4: Argue that because natural laws cannot be countermanded, women should resign themselves to remaining subordinate.

One famous example is the classification method for plants introduced by Carl Linnaeus in the eighteenth century. After counting the male stamens and female pistils in flowers, he replicated the hierarchy of Swedish society by first dividing them into twenty-four classes according to the number of male stamens and then relegating the female pistils to a subsidiary role within the main male groups. Once his taxonomic system had been adopted as orthodoxy, scientists could present it as a natural template on which human ranking should be based. Linnaeus's anthropomorphic interpretation of the world had been transformed into an inescapable fact of innate male superiority.

During the Victorian period, biological and medical ideas continued to reinforce the inferior status of women and make it more difficult for them to pursue scientific careers. The latest scientific theories of evolution and physiology worked hand in hand with long-established social convictions, strengthening the barriers preventing women from taking up professional positions in laboratories, universities, and hospitals.

Darwinian Differences

Charles Darwin exerted an enormous influence on attitudes towards women because his model of evolution provided an apparently rational justification of conventional Victorian beliefs. His basic model involves two stages, each repeated sporadically time after time. First an offspring happens to appear that is slightly different from its parents; next, that difference gives the offspring an advantage in the battle for survival in its immediate environment. Eventually, after repeated adaptations, a new species emerges that is better suited—more fit—for its surroundings.

Maintaining that men and women had diverged during evolutionary processes taking place over millennia, Darwin drove a wedge between the two halves of the human race.[3]

Darwin was by no means the first person to write about evolution; he was not even the first person to write about human evolution. His own grandfather Erasmus, who died before Charles was born, had published a long poem about the gradual development of living organisms from an initial 'ens' (living entity) up through insects, fish, mammals, and on up to human beings. Charles Darwin's innovation was natural selection as an agent of evolution. Although his model is now celebrated as a major scientific breakthrough, it aroused great controversy and was never fully accepted in its original form. Arguments still raged in the early twentieth century, and it was not until the 1930s that several approaches were melded together into a package resembling modern Darwinism.

Darwin's Victorian critics promptly pounced on several shortcomings. Most obviously, he had no way to prove that he was right. Despite piling up example after example in support of his ideas, Darwin could not explain why one generation should possess a new characteristic. What biological mechanism enabled two parents with identical flippers to produce a baby with flippers of a different style? He had plenty of circumstantial evidence, but no convincing explanation. He could not even point to an instance of evolution that was actually happening. Modern experimenters can simulate evolutionary pressures in a laboratory by breeding short-lived organisms such as beetles, but Darwin's system rested on hypotheses. His book is peppered with rhetorical question such as 'might it not be reasonable to suppose that...?' Very persuasive, but not the stuff of scientific proof.

Brooding for years over a theory he knew would be contentious, Darwin struggled to explain how some features might have conferred an advantage for survival. Human eyes, for example, were very problematic: how could such a complicated organ possibly have emerged in stages? Just thinking about it made him go cold all over, he told a friend. Even worse, the 'sight of a feather in a peacock's tail, whenever I gaze at it, makes me sick!'[4] How could it possibly be advantageous for a male bird to carry such a cumbersome tail? And why was the female so dowdy? To resolve this conundrum, he argued that the male's ostentatious display would allow him to pick the strongest and most fertile hens, a reproductive advantage that would outweigh the physical hindrance.

In 1859, Charles Darwin self-protectively refrained from mentioning human beings in *On the Origin of Species*, but by 1871 he felt ready to publish *The Descent of Man*, with its significant subtitle *Selection in Relation to Sex*. Moving from peacocks to people, Darwin claimed that equality was scientifically impossible. In line with his own ingrained assumptions, the standard ones of Victorian England, Darwin maintained that female inferiority is an inescapable consequence of nature. Men are cleverer, ran his argument, because over the millennia, their brains have become honed by chasing animals and defending their families. 'The chief distinction in the intellectual powers of the two sexes,' he wrote, 'is shown by man's attaining to a higher eminence in whatever he takes up, than can woman—whether requiring deep thought, reason, or imagination, or merely the use of the senses or the hands.'[5] To attract a powerful man, he explained, women compete by dressing elaborately. His admirers backed him up. A fashionable woman, commented H. G. Wells, surpassed even the extravagance of peacocks by providing 'an unwholesome stimulant' for men.[6]

Darwinian evolution implies a natural hierarchy, and was often interpreted to reinforce Victorian views on ethnicity as well as on gender. According to Darwin, just as lower creatures had evolved into higher ones, so too primitive races had evolved into more civilized ones. As these processes took place, the sexes diverged further and further, so that—he argued—male brains are superior to female ones; correspondingly, female characteristics such as intuition, empathy, and sensitivity are inescapable because they are biologically imprinted. For some of his followers, this meant that women are closer to animals and non-Europeans. As the sexologist Havelock Ellis put it, women are 'more curved forwards than the men', rather like apes and the 'savage races'.[7]

Conveniently, this scientific explanation supported domination not only over women at home, but also over the people of the British Empire. Although women everywhere were inferior, the problem was greater among the civilized races—so ran the argument—because divergence between the sexes had increased during evolution. Speaking in the House of Commons, an opponent of woman's suffrage argued that

> An adult white woman differs far more from a white man than a negress or pigmy woman from her equivalent male. The education, the mental disposition of a white or Asiatic woman reeks of sex; her modesty, her decorum, is not to ignore sex but to refine and put a point to it; her costume is clamorous with the distinctive elements of her form.[8]

These were not the words of the MP himself: he was quoting H. G. Wells, who reached wide audiences all over the country.

Measurements of brain weights and skull sizes seemed to corroborate views that white men (especially English ones) were the most highly evolved—that is, the best!—form of humanity. Expressing it the other way round, a London chemistry professor pronounced that because women were lower down the evolutionary scale than men, 'Education can do little to modify her nature'.[9] Even those sympathetic to science for women argued that they were better 'fitted' (that is, suited) to subjects such as chemistry or botany that required 'a capacity for noting details—patience and delicacy'.[10]

At the beginning of the twentieth century, noted a female physician bitterly, 'men tended to divide women into two camps. (1) Clever women and pretty women. (2) Good women and bad women.'[11] Presumably speaking from experience, she explained that in public debates, women often found themselves in a lose-lose situation. If they spoke in an appropriately feminine way, they were accused of being subjective and emotional—but if they argued rationally, they were warned not to overtax themselves, lest they endanger their health and their sanity.

However, women could take advantage of ambiguities within Darwinism. As originally formulated, evolution by natural selection is governed by chance rather than design, whereas Darwin's later theory of sexual selection suggests that people can direct evolution by choosing desirable partners. Striving to refute biological inevitability, suffragists denied that they were destined through evolution for marriage and motherhood. By placing the blame on social conditioning, they insisted that change was possible, that women could control the course of evolution by altering their behaviour. Cicely Hamilton, an outspoken journalist and playwright, argued that women had been schooled into submission so that they could fulfil their role in what she described as the economic trade-off known as marriage. In order to get her man, she explained, a woman learned to exaggerate the characteristics of passivity and stupidity that make her attractive as a bride: 'women have been trained to be unintelligent breeding-machines until they have become unintelligent breeding-machines.' But how short-sighted of men to adopt that strategy—suppressing women's intelligence would result in them suckling fools![12] Women needed to change their tactics, but she regretted that time would be needed to undo the damage: 'Think of the years, the generations, that women have been told they must not think! What wonder then that they make some mistakes when they begin to use the rusty instrument.'[13]

Another way of vindicating the suffragist cause was to claim that the modern woman occupied the next step up the evolutionary ladder. According to this argument, whereas men had long ago succumbed to animality by manifesting excess sexual desire, women were the civilizing influence who had clung to higher standards and could guide the future story of evolution by making the right choice of partner. The Woman's Freedom League suggested that in an industrial world, men were no longer adapted for superiority: technology in the home was a liberating force that would enable women to achieve their natural destiny of evolving still further. Modern inventions, the League argued, were making masculine strength redundant, so that large muscles would no longer be sufficient for attracting a choosy bride.

In this suffragist version of evolution, women would slowly gain ascendancy by educating society towards a higher state of morality. Adopting Darwinian terminology, campaigners maintained that 'the woman of political and social activity will be different from the domestic woman ... just as palaeolithic man differs from his neolithic brother'. Instead of being doll-like, the modern woman 'is now energetic and assured; not less beautiful, only differently beautiful'. To undermine the power of conventional arguments, they enlisted scientific vocabulary to castigate opponents as social dinosaurs. One cartoonist labelled an imaginary prehistoric creature the 'Antysuffragyst or Prejudicidon' (see Figure 3.1). Hampered by a tiny brain and sight so defective that it could not see past the end of its nose, it fed voraciously off the stupefying Humbugwort as it launched meandering attacks on its enemy, the female Justiceidon.[14]

STUDIES IN NATURAL HISTORY.
The Antysuffragyst or Prejudicidon.

The Antysuffragyst or Prejudicidon. This curious animal has the smallest brain capacity of any living creature. Its sight is so imperfect that it cannot see further than the end of its nose; but it has a wonderful capacity for discovering the stupefying plant called "Humbugwort," on which it feeds voraciously. It is closely allied to the Lunaticodon, and it is a fierce enemy of the Justiceidon.

Figure 3.1. 'The Antysuffragyst', *The Vote*, 26 September 1913.

Physiological Predestination

Darwin's theories were repeatedly contested and reinterpreted. Although arguments could be mutually inconsistent, one way and another the notion became steadily reinforced that men and women were inherently destined for different types of life. While many believed that women were physiologically suited only for virtuous domesticity, others worried that by linking human beings to animals, Darwin had made it permissible for base sexual instincts to take priority over higher human sentiments. Apparently on the assumption that English gentlemen always behaved with suitable restraint, this problem was seen as being particularly acute for non-Europeans and the lower classes; moreover, women could also lack the rational resources required to bring instinctive urges under control. What counted as appropriate sexual behaviour became a matter of scientific judgement. By decreeing what was normal, self-styled experts could also define deviation.

One particularly influential book, co-authored by the Scottish biologist Patrick Geddes and edited by Havelock Ellis, suggested that all sex differences stem from the fundamental distinction between small active sperm and large passive eggs. As a consequence, men and women possess opposite types of energy, which were given fancy Greek names—katabolic and anabolic. Men, ran the argument, are intrinsically vigorous and blessed with bigger brains, whereas women are weaker and more stable. Geddes is now celebrated as a pioneering city planner who designed Tel Aviv, but he also lent his name to the purportedly scientific argument that legislating for sexual equality would entail starting 'evolution over again on a new basis. What was decided among the prehistoric protozoa cannot be annulled by Act of Parliament.'[15] The laws of biology meant there was simply no point in giving women the vote.

Geddes and his colleagues were adding a modern scientific twist to old arguments. For centuries, women's supposedly erratic behaviour had been attributed to their reproductive systems. Ruled by their

bodies, women—often known as 'the Sex', as if in unspoken opposition to 'the Brain'—could be divided into two types: devoted wives
and mothers who found fulfilment in caring for their family; and
amoral prostitutes, those atavistic degenerates who quelled their
domestic instincts in order to sell their charms. The beautifully dressed
but militant suffragettes who stood at street corners selling *Votes for
Women* crossed categories. Men approached them as if to buy a paper
in support, but then whispered obscenities into their ears. During the
War, one of the strongest and most common accusations levelled
against working women was that they were cheap hussies looking for
sexual gratification. Men in khaki reportedly 'became in the eyes of a
number of foolish young women objects to be pestered with attention', and steps were taken to restrict women's movements in the
interest of protecting innocent soldiers.[16]

For many centuries, blame had fallen on the womb, whose Greek
name survives in the medical label 'hysteria', which until the First
World War was classed as an exclusively female disease. During the
second half of the nineteenth century, medical attention shifted
towards the ovaries. Seen as the counterpart to the male testes, they
were held to govern women's feminine characteristics by being controlled physiologically through the nervous system. By the time of the
War, researchers were learning about the importance of chemical
hormones, and would later shift onto them the responsibility for
women's behaviour. It is an 'essential fact', declared William Blair-
Bell, founder of the London College of Obstetricians and Gynaecologists, that 'femininity itself is dependent on all the internal
secretions....The female mind is specifically adapted to her more
protracted part in the perpetuation of the species.'[17]

One line of argument lay in portraying women as primarily maternal creatures who lacked the sexual desires exhibited by men.
Burdened as they were by the physical demands of menstruation,
childbearing, and lactation, women needed to lead quiet lives at
home. According to this script, in an ideal marriage the husband
acted not only as his wife's protector but also as the initiator of sexual

activity, which was designed for procreation rather than enjoyment.[18] Expounding this view, the London physician Arabella Kenealy insisted that nature prompted women to conserve energy for their vital task of reproduction. In her version of normality, women were inherently passionless creatures who could only become sexually aroused in direct response to stimulation from a man. In contrast, men were condemned to a life of violence in which they dissipated energy rather than conserving it. Men's intelligence had been refined by evolution, but 'the male sex-instinct may be seen still in all its native tyranny and selfishness ... the instinct manifests itself as an impulse of aggression, and the sex-function as one of brutality or ruthless lust.'[19]

Doctors articulated high-sounding versions of the 'frustrated spinster' argument. Walter Heape, the biologist who in 1890 had achieved the first successful embryo transfers (in rabbits), regarded unmarried women as 'waste products of our Female population', subject to madness because their reproductive organs had deteriorated through lack of use.[20] When suffrage demonstrations turned violent, critics diagnosed the problem as a medical one: in the absence of sexual activity, surplus energy was being released through unfeminine aggressiveness. Although sympathetic to the suffrage cause, the influential newspaper magnate Lord Northcliffe, pioneer of tabloid journalism, voiced this view discreetly. 'The war has proved that woman has not been given her opportunity in most parts of the Empire,' he declared in 1916; 'Today it begins to look as if the votes-for-women demonstrations were but manifestations of the tremendous pent-up energy of half the nation.'[21]

For the benefit of *Times* readers, the distinguished bacteriologist Sir Almroth Wright ('a fierce, hoary lion of a man who never spoke to a woman', according to one of his students) was more explicit. Relying on his reputation as an eminent physician, in 1912 he provided an explanation of female deviance couched in supposedly scientific terms. When 'the doctor lets his eyes rest on the militant suffragist,' he wrote, 'he cannot conceal from himself the physiological emergencies which lie behind' her mental disorder. 'These are the sexually embittered women in whom everything has turned into gall and bitterness of heart, and hatred of men.'[22] He advised shipping militant suffragettes off to the colonies, where there were plenty of spare men available for marriage.

For Better or for Worse

Evolutionary science might explain change, but it provided no guarantee of continued improvement. Perhaps the human race was evolving downwards—deteriorating? This was not a new fear: pessimists across Europe had been warning for decades that the human race was degenerating. Darwin's major publicist Thomas Henry Huxley insisted that 'it is an error to imagine that evolution signifies a constant tendency to increased perfection....Retrogressive is as practicable as progressive metamorphosis.' These biological warnings were endorsed by the latest theories in physics, which predicted that the universe would eventually run out of available energy. As the world ground to a standstill at some unspecified time in the future, only the very lowliest organisms would be able to adapt themselves for survival in the eternal winter of a cosmic wasteland.[23]

In Britain, gloomy prognoses focused in particular on the low moral standards said to prevail in city slums, where alcoholism was rife and poverty high. Recent events seemed to confirm the worst fears. The Boer Wars had finally ended in 1902, but their embarrassingly slow progress suggested that Britain might be losing its grip on the empire in Africa. In the first few years of the new century, army doctors declared that around half its recruits were unfit to fight; moreover, the number of mentally disabled people was apparently on the increase, and the birth rate was plummeting. The advent of war fuelled a type of patriotic puritanism among social reformers determined to cleanse the world and halt degeneration. In a chillingly savage sermon, the Bishop of London exhorted his congregation 'to kill Germans: to kill them not for the sake of killing, but to save the world; to kill the young men as well as the old ... to kill them lest the civilization of the world should itself be killed'.[24]

Advocates of compulsory conscription called on biological law, arguing that volunteers comprised 'the stout-hearted, the brave, the physically fit, the high-spirited, the patriotic': if they were to die, then those left behind would be 'the cowards, the physically unfit, the self-seeking'. Britain would have 'squandered its capital' of ideal fathers.[25] It seemed clear to many that if women abandoned their nurturing roles as mothers, the nation would decline still further. According to them, even if girls were sufficiently intelligent to come top of the mathematics examinations at Cambridge, their future duty lay in rearing clever boys, not indulging themselves in academic life. Maintaining the size of the population could not be entrusted to the lower classes if Britain were to safeguard its natural place as world leader.

Opponents of female education insisted that young women were ruining their health through the unnatural practice of studying. When a Manchester student died from tuberculosis, the Professor of Obstetrics had a one-word diagnosis: 'overeducation'.[26] In contrived couplets, *Punch* expressed another widespread worry—that overstuffed brains would produce masculinized spinsters:

> O pedants of these later days, who go on undiscerning,
> To overload a woman's brain and cram our girls with learning,
> You'll make a woman half a man, the souls of parents vexing,
> To find that all the gentle sex this process is unsexing.
> Leave one or two nice girls before the sex your system smothers,
> Or what on earth will poor men do for sweethearts, wives and mothers?[27]

In 1890, the Principal of Newnham, Eleanor Sidgwick, decided to test the claim that intellectual work made girls sick by carrying out a survey. Her conclusion was clear: she proved statistically that in fact her students' health improved rather than deteriorated during their three years at university. But even that solid scientific evidence did little to alter inbred prejudice. Darwin's cousin Francis Galton accepted Sidgwick's conclusion that studying could benefit young women, but twisted it to his own advantage. All the more reason, he argued, for graduates to marry early and have large families with a good hereditary pedigree. 'We want to swamp the produce of the proletariat by a better stock,' he wrote to Eleanor's husband Henry. In this informal letter to a male friend, he used the same vocabulary for training horses as for subduing women. 'It is a monstrous shame', Galton continued, 'to use any of these gifted girls for hack work, such as bread winning. It is as bad as using up the winners of the Oaks in harness work.' For him, the racing thoroughbreds of society's upper echelons should not waste their energy on mundane tasks of earning money—that was better left to the carthorses of the working classes. Generously, he offered to set up a fund for rewarding physically strong and morally impeccable young graduates. Prospective husbands, he warned, needed to be assured 'that marriage with such girls was *safe*', by which he meant that they would gladly renounce their books to focus on babies. As encouragement, he proposed awarding a prize to Newnham graduates first on their marriage and then on the birth of each child.[28]

One of Galton's legacies to the English language was the word 'eugenics', later notorious because it seemed to provide scientific justification of the Nazis' extermination policy. In the early twentieth century, eugenic strategies flourished in Britain, supported by many intelligent women as well as men who wanted to improve the British population. As the meanings of Darwinian terms slithered around, technical descriptions acquired emotive overtones. 'Fit' had originally referred to how well an organism matched its environment, but 'unfit' could easily be interpreted not as description but as condemnation. Because working-class families were often unhealthy and poorly educated, some reformers campaigned for family allowances, which

> would put an end to the increasing practice among all the more thrifty and farsighted parents of deliberately limiting the number of children, while the slum dwellers and mentally unfit continue to breed like rabbits, so that the national stock is recruited in increasing proportions from its least fit elements.[29]

But eugenics also had less charitable supporters who had little interest in improving conditions for the poor. One of the country's leading gynaecologists protested that to 'leave only the inferior women to perpetuate the species will do more to deteriorate the human race than all the individual victories at Girton will do to benefit it'.[30] He knew that by 'inferior', many people would immediately understand prostitutes and the lower classes (especially the Irish), who were deemed nearer to animals and hence governed by their sexual urges.

H. G. Wells entered the eugenics arena less aggressively. In his imaginary vision of a perfect society, he insisted that women should be paid a salary to fulfil their major function in life—bringing up healthy children to benefit the state. The more highly achieving the children, the higher the maternal wage would be. In his ideal state, women would remain faithful to their husbands (although men would be allowed to roam), and childless marriages could be terminated. 'In Utopia a career of wholesome motherhood would be...the normal and remunerative calling for a woman', wrote Wells; 'a capable woman who has borne, bred and begun the education of eight or nine well-built, intelligent and successful sons and daughters would be an extremely prosperous woman, quite irrespective of the economic fortunes of the man she has married.'[31]

Worried about the future of the British nation, Arabella Kenealy— not only a doctor but also the author of racy lesbian novels such as *Dr Janet of Harley Street*—argued that women were much better than men at assessing suitable partners. If only others would follow her own example by spurning marriage to become economically independent, she argued, then those women who actively wanted a husband would have more choice and could select better fathers for their children. As if applying the laws of thermodynamics to human beings, she asserted that 'it was a crime against the next generation, for women who hope some day to be mothers, to spend in study or labour the physical and nervous vitality which should be stored up as a kind of natural banking account' to ensure that any future children would be strong and healthy contributors to British greatness.[32]

Unsurprisingly, many suffragists saw the situation rather differently. Even so, they did not automatically attack eugenic initiatives. The paleo-botanist Marie Stopes, who marched with the militant suffragettes, joined the Eugenics Society in 1912 and six years later published *Married Love*, which included basic information about birth control. Her aim was similar to Galton's—she wanted to see British people grow in strength and beauty. But whereas Galton thought privileged young women should abandon any hopes of a career and instead stay at home to produce babies, Stopes encouraged inferior women (her terminology, not mine) to learn the contraceptive techniques that were already familiar to the middle classes.[33]

Scientific ideas were deployed to back up opposition to the suffragist campaigns, and neither humour, violence, nor rational argument made any substantial dent in received wisdom. Endorsed by many scientists and doctors, the majority view remained essentially unchanged: women were born to reproduce, and any attempt to educate them would disturb their natural constitution. In a parliamentary debate of 1910, the MP Austen Chamberlain explained that he opposed giving the vote to women because they were inherently different from men. 'In my opinion,' he declared, 'the sex of a woman is a disqualification in fact, and we had better continue to so regard it in law.'[34] It was not until 1928 that all British citizens over twenty-one gained equal voting rights.

PART II

ABANDONING DOMESTICITY, WORKING FOR THE VOTE

4

A NEW CENTURY

Voting for Science

Does history repeat itself, the first time as tragedy, the second time as farce? No, that's too grand, too considered a process. History just burps, and we taste again that raw-onion sandwich it swallowed centuries ago.

Julian Barnes, *A History of the World in 10½ Chapters* (1989)

Whatever Karl Marx might have said about the task of philosophers being to change the world, the suffragists of the early twentieth century knew that great thoughts alone could not overturn assumptions that had built up over generations. 'Deeds not words!' ran a slogan of the militant suffragettes, and even peaceful campaigners laid on increasingly spectacular events. Muriel Matters, an Australian actress with a talent for getting herself noticed, flew up in an air balloon with 'Votes for Women' painted in large letters on its side (see Figure 4.1). From her seat in the cane basket suspended below, she distributed thousands of leaflets that fluttered over the landscape, urging the nation to support the suffrage movement.[1]

To take full advantage of the new cheap newspapers with their rocketing sales figures, these media-savvy women ensured that journalists were well briefed. Recruiting camera crews, they projected films onto sheets hung in shop windows, thus advertising not only their marches but also their familiarity with the very latest technology. By the time of the First World War, there were almost 4,500 cinemas in Britain, each showing a newsreel for several days before passing it on to the next. Unfortunately, very few of these old films survive, but

Figure 4.1. 'Votes for Women' blimp, *Daily Mirror*, February 1909.

suffragists made sure that their activities featured regularly. Emily
Wilding Davison might not have intended to die when she leapt in
front of a horse at the annual Derby held at Epsom, but she knew that
press photographers were already swarming around the racecourse—
and can it be coincidence that she chose Tattenham Corner, right in
front of a mounted camera that would be sure to capture her dramatic
self-sacrifice?[2]

Figure 4.2. Emily Jane Harding, 'Convicts Lunatics and Women!', c.1908.

When they marched through the streets, suffragists forced traffic to a halt—then a newsworthy event in itself—and waved beautifully crafted silk banners, many of which are still lovingly preserved in museum archives. Adapting traditional imagery from the Trades Unions, teams of embroiderers celebrated the careers that women were starting to enter.[3] One particularly striking image of a graduate was designed by Emily Harding, a children's book illustrator too old and of the wrong class to have gone to university herself. Her yellow and black banner showed a refined young woman wearing a mortar board and full academic dress, but trapped behind a high locked gate with two men physiognomically depicted as 'degenerates' (see Figure 4.2). 'Convicts Lunatics and Women! Have no vote for Parliament', says the caption—should all women be categorized thus?[4] Even those women sufficiently clever and privileged to gain a degree were not judged to be responsible adults.

Being Modern

Historians segment the past into centuries, as if the switch from '99 to '00 automatically heralded a new era. But of course, for people at the time, 1900 felt little different from 1899, just as on the arrival of 2000, everyday life continued as normal despite the panic about a millennium bug. By the time of the War, older people had come to terms with living in the twentieth century rather than the nineteenth. In contrast, the young men going off to fight could scarcely remember the Victorian era, while their sisters regarded themselves as modern women, very different from their *fin-de-siècle* mothers, who had in turn looked back askance at the previous generation.[5]

Hedda Gabler, the manipulative central character of Henrik Ibsen's domestic drama, had first appeared on a London stage in 1889. Storming out of her repressive marriage to pursue an independent existence, she rapidly became a figurehead for real-life British women who claimed the right to vote and to run their own affairs. Conventional ideologies had been under threat for decades. The Victorian ideal woman had been colloquially dubbed 'the angel in the house' after the title of a popular sentimental poem by Coventry Patmore that hymned the perfect wife who devoted herself to her husband and children. Before the end of the century, that particular domestic goddess had been more or less banished. Her iconic place was taken by the 'New Woman', a catchy label that flourished in books, plays, and articles to denote anyone who defied traditional stereotypes of domestic femininity and subordination. This liberated female who thought for herself and dressed as she liked was often satirized—a sure indication that she posed a real threat.

Edwardian suffragists demanded to be taken seriously as rational human beings with no need to shelter behind a mask of fragile frivolity. Hedda Gabler no longer seemed an ideal role model— emotionally unstable, she was too aware of her own sexuality and focused on her personal problems. Rather than seeking individual escape from their own trapped lives, campaigners wanted equality and freedom for all women. After working-class women became involved in the movement (although only rarely in its leadership), suffragists broadened their agenda. Although, like their Victorian predecessors, they still wanted to remedy the plight of independent ladylike spinsters, they also began agitating about industrial pay and conditions, childcare, and marital rights. Winning the vote remained important, but for those without a wealthy, indulgent family, financial independence and parity seemed more urgent.

Modern women, declared Emmeline Pankhurst's American ghost writer, demanded 'to belong to the human race, not to the ladies' aid society to the human race'.[6] Her pithy sound bite gave neat expression to an old sentiment. Women had been campaigning for equality since the middle of the nineteenth century. The First World War may now seem a dramatic hiatus, but as perceived by women living through their own experiences at the time, the present appeared to be developing continuously from the past. Unaware of what lay ahead, they were building upon the reforms and the organizations that they had inherited.[7]

The British nation had unofficially moved towards militarization well before the War—the Girl Guide motto 'Be Prepared!' evidently expressed a common sentiment. Parents enrolled their children in hierarchical organizations modelled on the Army, such as the Boy's Brigade and Robert Baden-Powell's Scouting movement. Initially, both boys and girls were encouraged to wear uniforms, practise leadership, and engage in physical exercise, but before long, the girls were being channelled into what seemed more appropriate activities, such as first aid or sewing. Rebranded as Girl Guides and consigned to the leadership of Robert's sister Agnes, female scouts were no longer allowed to enjoy the same challenges as their brothers. For them, excitement was reduced to vicarious experience: the girls were instructed that in the event of a battle, they should 'be plucky and go out and do something to help your brothers and fathers who are fighting and falling on your behalf'.[8]

During the War, Baden-Powell's wife Olave took over from Agnes and adopted a policing approach to her role of Chief Commissioner. Dismissing old-fashioned homilies about the value of lace-making, she exhorted her Guides to prepare sandbags and collect sphagnum moss for bandages. One among many self-appointed elite guardians of the nation's morals, and apparently convinced that hordes of heavily made-up nymphomaniacs were enticing young soldiers in uniform, Olave imposed an orderly regime. Focusing on clean living and the outdoor life, she warned about the 'brazen flappers' who pretended to do war work, but were actually seducing men. When a girl 'has learnt the ABC of sin', she intoned, 'it is but a step towards the first primer of vice'.[9]

Edwardian women emulated their menfolk by marshalling themselves into miniature armies of activists. Founded in 1907, the First Aid Nursing Yeomanry (FANY) recruited young ladies from privileged families who could pay a monthly subscription, buy a smart red, white, and blue uniform (later changed to khaki), and knew how to ride (preferably providing their own horses). The most famous is Flora Sandes, who was eventually promoted to captain in the Serbian Army and made sure she was always photographed in her military uniform.

Later, after the War started and they were called into action, some disillusioned members pined for their comfortable fireside sofas. 'And oh!' complained one volunteer FANY driver—dubbed a 'butterfly chauffeuse' by an unsympathetic friend—'to be able to wear a crêpe de Chine blouse instead of this everlasting manly uniform!'[10]

In contrast, more adventurous teenagers relished the opportunity of rebelling against their parents by donning overalls instead of a debutante's ballgown:

> I wish my mother could see me now
> With a grease gun under my car
> Filling the differential
> 'Ere I start for the sea afar...
> I used to be in society once
> Danced and hunted and flirted once
> Had white hands and complexion once
> Now I am FANY...[11]

Modernity was often associated with science. In the early twentieth century, inventions were proliferating and enthusiasm was growing. 'I think we were all a little mad', recalled the aviation pioneer John Moore-Brabazon; 'we were all suffering from dreams of such a wonderful future'. Discovered in 1895, X-rays signalled by their very name that unknown and unsuspected forces were at work. Before long, women could buy radioactive face creams and hair tonic. Some of them supposedly relied on the imaginary substance 'Radio-Pilocarpine' (see Figure 4.3), while others were 'guaranteed to be the finest therapeutic preparation known to science [containing] the correct quantity of *Actual Radium*'.[12] Long-distance communication by radio was becoming a reality, the theory of special relativity appeared in 1905, and in 1911 Ernest Rutherford discovered the atomic nucleus. The British Army is often castigated for maintaining reactionary attitudes, but it too was making itself scientific. In 1905, a lawyer called Richard Burton Haldane was appointed as Secretary of State for War. Influenced by German philosophy, he believed—with a passion verging on faith—that life should be guided by rational principles. For

Figure 4.3. Advertisement for
Nelson's Radio-Pilocarpine hair tonic.

Haldane, science was the only guarantor of progress, and he set about
reforming military technology and diverting government resources
towards aviation, still in its infancy.[13]

To demonstrate their modernity, suffrage supporters allied themselves with scientific and technological progress. The book popularly

known as the 'Bible of the Woman's Movement' was *Women and Labour*, published in 1911 by the South African pacifist and suffragist Olive Schreiner, now best known for her novel *The Story of an African Farm* (1883). '[T]here is no fruit in the garden of knowledge it is not our determination to eat,' Schreiner proclaimed; 'from the chemist's laboratory to the astronomer's tower, there is no post or form of toil for which it is not our intention to fit ourselves.'[14] Writing both as a suffragist and a committed socialist, Sylvia Pankhurst (Emmeline's daughter) saw technology as the route to emancipation for all human beings. How strange, she commented cynically, that so many of her suffrage friends defended the right of women to toil like men—surely they should be campaigning for investment in machines that would reduce their workload? If they visited the industrial Midlands, she claimed, middle-class suffragists would see for themselves that 'women dragged and pushed the heavy wagons of coal, others picked the rubbish from the fuel as it passed on moving belts—all work which invention must reduce, and, if possible, eliminate for the liberation of human souls.'[15]

Both suspect activities for women, science and suffrage were often bracketed together by critics worried about this twin threat to long-established conventions of female domesticity, subservience, and demureness. In *Ann Veronica*, H. G. Wells satirized this nationwide mix of fear and aversion through the reactionary father of an independent daughter. Blind to her ambitions and frustration, this endearingly realistic yet blinkered father finds it impossible to understand why she should be 'discontented with her beautiful, safe, and sheltering home, going about with hatless friends to Socialist meetings and art class dances, and displaying a disposition to carry her scientific ambitions to unwomanly lengths'.[16] The rows between Ann Veronica and her parents replicated angry discussions around dinner tables in homes all over the country. What family scenes were played out in 1907, when the chemistry graduate Alexandra Wright abandoned her study of bees and devoted herself instead to the suffrage movement? The Professor of Forensic Medicine at Aberdeen was presumably mortified when his daughter Helen Ogston, a science graduate and trained sanitary inspector, was featured in the papers after lashing out with a dog-whip at a women's meeting addressed by Lloyd George in the Albert Hall.[17]

Scientific and technological innovations were central to feminism, and the relationship operated in both directions. New inventions affected how women behaved—and conversely, attitudes towards women affected how technology developed. Take bicycles. First introduced in the early nineteenth century, they were initially unstable and impossible to ride in long, flowing skirts. In response to women's demands for safer machines, during the 1890s new devices—brakes, gears, air-filled tyres—were introduced to make cycling easier and more comfortable. Adventurous women were quick to jump on the saddle, but inevitably, reactionary critics were equally quick to object. According to them, such behaviour was not only undignified, but also threatened that most important aspect of a woman—her ability to bear healthy children.[18]

As ever, appearance was a major issue, and the shorter skirts, trousers, and loose clothes adopted by defiant cyclists sparked controversy. Speaking in their defence, progressive doctors warned about the dangers of wearing tightly laced corsets, and at an annual meeting of the British Association for the Advancement of Science, scientists were given a lecture advocating 'Rational Dress' by the self-educated militant suffragette Charlotte Stopes (Marie's mother).[19] But traditionalists continued to sneer. When Bertrand Russell's first wife ventured out in her own version of sensible clothing, a comfortable if unconventional costume of 'loose Turkish trousers, Zouave jacket and broad sash round my waist', her aristocratic neighbour was appalled by such a lack of decorum. Declining an invitation to tea, she wrote sniffily: 'Since I met you bicycling in the Lodsworth Road, it has been impossible for me to consider you as friend.'[20]

Such objections stimulated further technological changes. Seeking to broaden their market, ingenious manufacturers introduced another adaptation: wheel-guards, which protected dresses against becoming tangled up in pedals and spokes, and so enabled more conservative women to ride safely. Gradually—and as so often happens—what had once been outrageous came to be accepted as the norm. As early as 1895, the *Manchester Guardian* explained that 'thanks to the illustrated papers and the recent developments of gymnastics and cycling, the general public has become so familiar with "Rational Dress" that it no longer creates anything like a sensation.'[21] To refute suggestions that frequent breakdowns and punctures prevented women from cycling on their own, female experts published detailed instructions on carrying out repairs. After all, these authors argued, needles and scissors are just as much tools as screwdrivers and spanners.

Formerly mocked as inappropriate, cycling became accepted as a healthy form of outdoor exercise for everybody. It also helped to give modern women unprecedented freedom and independence. Similar debates arose when cars were introduced. Initially, it was deemed improper for ladies to take the wheel: after all, like women and horses, cars needed to be controlled by a man, and they were generally referred to as 'she'. Once women started earning money, manufacturers expanded the potential market by making cars easier to drive, providing lighter brakes and electric starters to replace unwieldy mechanical cranks. Even so, they often aimed their advertising slogans towards male purchasers, targeting husbands wanting to buy a treat for their wives, or businessmen hoping to hire a female driver.[22]

Science was also changing women's lives at home. Before the War, wealthy women relied as previously on hard-working, underpaid servants to do all their washing, cleaning, and cooking. But now they were being bombarded with advertisements for commercial products promising to enhance their lives and reduce their workload—carpet cleaners, ovens, lavatories, boilers, telephones, wind-up phonographs, processed food. Some enterprising British women were patenting their own home improvements—hand-operated washing machines, two-part moulds for steaming puddings, piston-operated mustard pots, extendable beds, air-filtering roller blinds, telescopic window-cleaners.[23]

These new technologies affected different homes in different ways. Most purchasers belonged to the middle classes: poorer families could not afford the latest labour-saving devices, while affluent ones saw little point in reducing the responsibilities of their staff. Domestic inventions proliferated after the War, when many former servants refused to resume their below-stairs existence following their experiences of more lucrative work in factories and offices. As the burden of housework shifted up the social scale, automated cookers and cleaners could paradoxically result in *extra* work for housewives trapped at home, who were expected to justify their expensive purchases by serving elaborate meals and maintaining ever-higher standards of cleanliness.

Similarly, although new technology transformed office life, it brought unforeseen consequences. Typewriters benefited women, enabling them to gain employment by replacing Victorian male clerks with their impeccable copperplate handwriting. Yet perversely, the new machines helped to steer women into low-paid, low-status jobs. Typists were offered low wages, which meant that only women would accept the position—and as a consequence, typing became devalued as an essentially female preserve. Unsurprisingly, many women preferred working in an office to cleaning someone else's floors, and they were willing to take up clerical jobs at lower rates than their male competitors. Soon they were relegated to being the inferior assistants of men. Whereas their previous menial tasks had involved sorting out an employer's laundry and arranging china on shelves, now they were tidying up sloppy grammar and spelling. As suffrage campaigners protested, this type of cycle was often repeated. However liberating technological innovations might have seemed at first, they did not necessarily lead to long-term benefits.

Demanding the Vote

Although suffrage campaigners prided themselves on being modern, they borrowed from the past when it suited them. 'They have recreated the beauty of blown silk and tossing embroidery,' reported a journalist admiring a march on a windswept day in February 1908; 'The procession was like a mediaeval festival, vivid with simple grandeur, alive with an ancient dignity.'[24] Three years later, on the day before George V's coronation, 40,000 suffragists—including graduates and doctors—trouped along the already decorated route. As the *Manchester Guardian* reported,

> The Oxford banner, to pick one from the list, was exquisite with its representation in gold of spires and domes....Ahead rode Joan of Arc in silver armour followed by a symbolic group of New Crusaders, wearing purple mantles, women crowned as queens but servants at the same time, for each queen carried the train of the queen in front.[25]

For decades before international warfare broke out, British suffragists had been engaged in their own internal fights. Any apparent surface unity concealed a reiterating turmoil of concealed schisms, with temporary periods of peace being succeeded by fresh outbreaks of rivalry. In 1897, twenty disparate groups attempted to strengthen women's collective voice by coalescing to form the National Union of Women's Suffrage Societies (NUWSS), led by Millicent Fawcett and dominated by energetic middle-class women. There followed what one author described as 'the flowering time', when meetings were conducted with missionary fervour to draw in ever-widening circles of enthusiasts.[26]

But within a few years there was a major disagreement about strategy. Rueing the lack of progress made by polite meetings and marches, some women insisted that it was time to adopt more aggressive tactics. In 1906, the *Daily Mail* coined a new word to describe militant activists—suffragettes. Less democratic and well-organized than the NUWSS, they regarded themselves as an army of volunteers unhampered by legal restrictions. Nicknamed 'The General', Flora Drummond rode in front of her female troops wearing an officer's cap and epaulettes—but unlike a conventional leader, she was imprisoned nine times for protests that included demonstrations inside the House of Commons and at 10 Downing Street. Convinced that they could force through change where constitutional means had failed, these women obliged the nation to take notice of their demands by chaining themselves to railings, heckling meetings, and going on hunger strike in prison.

The NUWSS remained the largest organization, followed by the Women's Social and Political Union (WSPU), which was founded in 1903 by Emmeline Pankhurst, widow of a Manchester barrister. It soon splintered into smaller rival sects; although Emmeline's daughter Christabel remained loyal, Adela emigrated (financed by her furious mother), and Sylvia deserted the WSPU to take up the socialist cause. The members of this quasi-military organization came to regard themselves as spiritual soldiers, displaying increasing levels of violence yet regarding themselves as morally guiltless whatever crimes they committed for their cause. Divided into contingents for marching in processions, they wore stylish uniforms and draped themselves in distinctive purple (dignity and loyalty), white (purity), and green (hope) sashes—even their umbrellas were colour-coordinated.[27]

The suffragettes certainly succeeded in drawing attention to their cause, but many people—women as well as men—were appalled by their extreme behaviour. Exposed to public ridicule and humiliation, some activists soon abandoned militancy, deploring what they perceived as a herd-like adoption of unthinking violence inspired by the Pankhursts among their followers.[28] One female journalist was struck by the level of class and gender hatred she observed. 'The men's faces were inflamed and distorted,' she reported; 'The opportunity to manhandle educated and defenceless women of refined birth and upbringing appeared to them as a particularly stimulating and unusual form of "rag".' In contrast, the women remained serene, glowing with an inner light as they ignored the 'howls and blows and bestial faces' surrounding them.[29]

Some campaigners went on to become pillars of the establishment. Decades later, after the Second World War, the head of the Women's Royal Naval Service met the Labour Prime Minister Clement Attlee. 'When, at the age of twenty,' she told him, 'you have stood in the gutter selling the *Suffragette* while passers-by spat on you, at the age of fifty you can face anything!' Attlee knew exactly what she meant. 'I was very young and very shy,' he replied, 'when I stood as one of four in the streets of Limehouse singing the "Red Flag" to collect a crowd for a meeting. That, too, is an experience which helps you to face anything that may come in later life.'[30]

Aware that they were regarded with suspicion, even non-militant suffragists had to tread carefully. Wary of jeopardizing their fragile status, female scientists tended to opt for the more diplomatic NUWSS. At Newnham, the cricket captain Ray Costelloe initially found the excitement of militant action appealing, but she became a staunch opponent. 'I wish I was a militant,' she confided to her diary in 1910, 'but one *can't*! They are so repulsive as well as so fine!'[31] Like many NUWSS members, her resolve to act peacefully was challenged the following year. In May, on a second attempt, there was an enormous majority in Parliament for a bill that would extend voting rights for both men and women. By November, Prime Minister Asquith had yielded to pressure and set a date for a Male Franchise Bill, while remaining provocatively silent about female franchise. Nearly two decades later (now writing as Ray Strachey), she still remembered the 'feeling of bitterness, of betrayal, and of political disillusionment' that tempted law-abiding women to join the militants, whose protests became more virulent than ever.[32]

Alarmed at the bad publicity generated by the militant suffragettes, the NUWSS successfully attracted reporters by adopting a different advertising strategy: holding peaceful demonstrations. Identifying themselves with the symbolic colours red, white, and green, the NUWSS underlined their non-militancy by featuring a bugler girl on posters and other promotional material: 'her sword is sheathed by her side; it is there but not drawn, and if it were drawn, it would not be the sword of the flesh, but of the spirit. For ours is not a warfare against men, but against evil' (see Figure 4.4).[33] They also emphasized their non-aggressive femininity by embroidering beautiful banners, featuring the bugler girl and other symbols. The message was clear. Women

Figure 4.4. Caroline Watts, 'Bugler Girl', 1908.

could think, hold down responsible jobs, and march for their rights, but they could also sew and behave like real women.

The number of women who joined the NUWSS protesters was newsworthy in itself. Unfortunately, the first major event, in 1907, was in February, timed to correspond with the opening of Parliament but not a propitious month for a mass demonstration. Even so, over 3,000 women trudged through the streets of London on a cold, wet day, carrying banners and accompanied by several brass bands. The streets were lined with spectators, and on the unpaved roads, the women's shoes and ankle-length dresses became so clogged with mud that the event became known as Mud March. In papers all over Europe and America, journalists marvelled at both the size and the diversity of the procession. Women of all ages, classes, and professions braved not only the weather, but also public ridicule (to say nothing of the reaction from their menfolk at home). It was only afterwards that they realized how seriously the crowd had taken them.

Mud March was big, but the following year an even bigger demonstration took place in June (a far more sensible choice than February). Provoked by the recent blocking (one among several) of a suffrage bill in Parliament, over 10,000 women processed through London to a NUWSS rally in the Albert Hall. By then, a leading Arts and Crafts designer called Mary Lowndes had recruited a miniature army of embroiderers to create stunning banners, some identifying regional participators from all over the country, others referring to professional occupations, and eventually including scientists such as Mary Somerville, Caroline Herschel, and Florence Nightingale. Figure 4.5 shows the design for a suffrage tribute to Marie Curie, later hand-crafted in striking orange and yellow silk fabrics. Her initiative demanded skills of diplomacy as well as needlecraft. 'Oh! My odious Committee!!' she moaned to a colleague; 'They were here till nearly 11 last night—cross, quarrelsome, undecided and prolix . . . an *awful* team to drive.'[34]

A still more impressive procession took place in 1911, when 40,000 demonstrators from all over the world marched five abreast in a line

Figure 4.5. Design for Marie Curie
Radium banner.

that stretched for seven miles. The crowd's greatest applause was for
working women, especially nurses, but there were also academic
representatives. 'Whitehall Gardens, you found at this time possessed
by some hundreds of the women graduates waiting their turn,'
reported an admiring journalist; 'Women in scarlet doctors' robes
sat on the steps of the deserted Government offices.... The stately
band of university women made a deep impression... the college
banners made a good show.'[35]

As the government continued to prevaricate, tension grew. While
the NUWSS focused on recruiting members and fine-tuning its policy,
Pankhurst and her followers collected large sums of money to finance
their increasingly outrageous exploits. Aiming at the affluent, they
marketed button badges, colour-coded underwear, and even shoes
patterned in the suffragettes' purple, white, and green. The situation
exploded in 1912 when, on the third attempt to get a bill for voting
reform through Parliament, the government seemed to renege on its
promises of the previous year. Despite claims that a legal technicality
was to blame, suffrage leaders were in no doubt where the guilt lay

and they said so: either Prime Minister Asquith had made a stupid mistake, or else he had deliberately cheated women out of an opportunity to win the vote. Militant suffragettes stepped up their activities, grabbing the headlines by attacking policemen, damaging property, and releasing gruesome accounts of being force-fed in prison. In March 1912, several hundred swarmed through London streets, smashing in shop windows with hammers concealed in parcels or handbags. Sited uncomfortably near Harrods, the Natural History Museum closed its doors for three weeks.[36]

In retaliation against the repeated acts of violence, Emmeline Pankhurst was imprisoned and in 1913, the notorious Cat and Mouse Act was passed. This prevented women from gaining publicity by starving themselves to death: in a repetitive cycle, as soon as a prisoner on hunger strike reached the limits of her endurance, she was released but then rearrested as soon as she had recovered some of her strength—like a cat teasing its victim. Among the eminent women lending support to the suffragettes was the physicist Hertha Ayrton, who sheltered Emmeline Pankhurst during a temporary release from jail and contrived to help her elude recapture by sending out a decoy carriage when the house was besieged by police and journalists. Ayrton was a close friend of Marie Curie and taught mathematics to her daughter Irène. Although Curie usually withheld her name from petitions, Ayrton persuaded her through Irène to sign a protest against suffragette imprisonment.[37]

Although still rejecting violence, the NUWSS also stepped up its activities. In June 1913, they launched what became fondly known as 'The Great Pilgrimage'. For six weeks, troupes of marching women converged on London from all over the country. Dressed mainly in sober grey, white, and black uniforms, they swept up allegiants along the way until the crowds numbered around 50,000 for the final peaceful rally in Hyde Park. Even the Prime Minister conceded that these well-organized protesters deserved some claim on his attention.

In contrast, militant suffragettes risked damaging their own cause by antagonizing the public—other women as well as men. Violence intensified during the first half of 1914, and when pictures were slashed, museums and art galleries introduced time-consuming and aggravating bag and muff checks, the equivalent of airport security controls. After a protester attacked a man with a horsewhip at Euston Station, an assassination attempt began to seem a realistic possibility, and during the hot summer months suffragette leaders found it hard to control their members. These apparently pointless attacks by individuals threatened to confirm the widespread opinion that women were emotional, irrational creatures who did not deserve the vote. During the week of heavy rain leading up to the August bank holiday of 1914, resentment deepened.[38]

When war was declared, Emmeline Pankhurst's daughter Christabel took immediate advantage of the crisis, declaring that 'This great war...is Nature's vengeance—God's vengeance upon the people who held women in subjection.... Only by the help of women as citizens can the World be saved.'[39] Many campaigners—suffragettes as well as suffragists—publicly declared an immediate if temporary truce, although some lobbying still continued behind the scenes. Looking back, Newnham's former cricket captain, Ray Strachey, glorified this apparently overnight transformation. 'The impulse of violent patriotism burned in their hearts, and the longing to be of service moved them,' she reminisced in 1928; 'They knew they were able and energetic; they knew that they had gifts to offer and help to give to their country; and rich and poor, young and old, spent their time in searching for useful work.'[40]

Like so many retrospective reflections, this was a romanticized black-and-white version of messy reality.[41] Even so, countless campaigners rebranded themselves by discarding their previous identities to don workers' overalls, or army uniforms in sober khaki with a splash of patriotic red, white, and blue. No longer marching for votes, these women committed themselves to working for war.

5

FACTORIES OF SCIENCE

Women Work for War

Violet Velma Vere de Vere
Was a dainty, delicious, delicate dear;
Delightfully dressed from top to toe,
For she always 'got it at VENN'S, you know!'

Exquisite, filmy, lacey things,
(My imagination, you see, has wings!)
Were Violet's dearest dream of delight
By day and—whisper it gently!—night.

Now the very last thing you'd have thought she'd do
Was to 'work on the land.' But she did! It's true!
For she couldn't nurse and she wouldn't knit,
But was awfully keen on 'doing her bit.'

Never a day but she rakes and digs,
And plants potatoes and feeds the pigs...
But doesn't she just enjoy her Sundays,
When she revels again in her dainty 'undies'!

Advertisement for Venn's Undies,
Tatler, 2 May 1917

Caroline Haslett was just one among many thousands of young women whose lives were transformed by the First World War. Through their struggles, setbacks, and successes, they collectively influenced future generations. Her experiences illustrate how the War permanently altered scientific, medical, and technological prospects for women. A suffragette with a weak school record, she became an eminent international consultant on the domestic uses of electricity, educational reform, and industrial careers for women. She used her influence to alter the scientific careers of countless schoolgirls all over the world.

Haslett was judged a lost cause by her teachers because she never could learn how to sew a buttonhole. As a teenager, she left her Sussex village for London and—to the alarm of her strict Protestant parents—joined Emmeline Pankhurst's suffragettes. When the War started, she was working as a clerk in a boiler factory, but during the next four years she was repeatedly promoted to replace men who had left to fight. By 1918, she was running the London office, visiting customers such as the War Office to discuss contracts, and astonishing staid civil servants with her expertise in a man's domain. After being trained as an engineer by her enlightened employers, in 1919—still in her early twenties—Haslett began managing the newly founded Women's Engineering Society. She was determined to consolidate and expand still further the opportunities for women that had recently opened up. Electric technology—dishwashers, vacuum cleaners, washing machines—would, she believed, free women of drudgery, liberating them to lead a higher form of life. She envisaged

> a new world of mechanics, of the application of scientific methods to daily tasks ... a great opportunity for women to free themselves from the shackles of the past and to enter into a new heritage made possible by the gifts of nature which Science has opened up to us.[1]

At the end of the War, three million women were working in industry. Like Haslett, some of them had the advantages of good grammar and the right sort of accent, but many were relatively uneducated—domestic servants, barmaids, and shop assistants who had seized the opportunity to escape from their menial occupations. Critics savagely denounced them either for abandoning their natural sex to behave like men, or for being spendthrift prostitutes. Overcoming the nation's reservations, these hard-working wage earners proved their capabilities. By 1917, even Herbert Asquith, long a fierce opponent of female suffrage, had been converted. 'How could we have carried on the War without them . . . ?' he asked; 'Short of actually bearing arms in the field, there is hardly a service . . . in which women have not been as least as active and as efficient as men.'[2]

Would women have won the vote in 1918 without these dedicated wartime workers? Why was the vote given to impoverished men but not to affluent young women? Did the militancy of the suffragettes harm or hinder their cause? Those remain some of the unanswerable what-if questions of history.[3] What does seem clear is that public opinion was nudged towards approving female suffrage in recognition of women's wartime contributions. In 1917, the Prime Minister David Lloyd George told Parliament: 'I myself, as I believe many others, no longer regard the woman suffrage question from the standpoint we occupied before the war. . . . The Women's Cause in England now presents an unanswerable case.' Addressing a less select audience, the *Daily Mail* delivered the same message: 'The old argument against giving women the franchise was that they were useless in war. But we have found out that we could not carry on the war without them.'[4]

Women at War

According to Lloyd George, Chancellor of the Exchequer when War broke out, Europe unintentionally and unwittingly slithered over the brink into a boiling cauldron. The government declared it would be a short-term conflict taking place on the other side of the Channel, and at the end of 1914, Winston Churchill reassured the nation that its motto should be 'Business carried on as usual during alterations on the map of Europe'. Concealed behind this official rhetoric, strategic measures had long ago been set in place. Planning ahead, Churchill had already insisted that money should be poured into modernizing the Navy and buying the latest technological equipment. By using the power of this new naval science, Britain aimed to starve Germany of supplies but keep the sea lanes open so that allies could be supplied with munitions.[5]

Because German imports were abruptly cut off, Britain found itself running out of essential scientific supplies—electric generators, tungsten, explosives, medical drugs, high-quality glass for range-finders. Soldiers repeatedly complained that the shortage of shells prevented them from fighting effectively, and publicity intensified after Lord Northcliffe, influential owner of the *Times* and the *Daily Mail*, accused the Army of causing the death in action of his nephew by not supplying sufficient ammunition. To resolve the situation, in 1915 Lloyd George was put in charge of a new Ministry of Munitions, which started establishing factories all over the country, operated largely by women.

Many organizations responded rapidly when war was declared in August 1914. For advertisers, the emergency represented a marketing opportunity: enterprising advertisers repackaged existing products as 'Active Service Kits', designing romantic scenes of glamorous young women waving off their men with packets of cigarettes.[6] Suffrage groups were also quick to recognize the good publicity that might be generated by switching from protest to patriotism—which is not to minimize the enormous and genuine surge of national loyalty that inspired many women. Only two days after hostilities were opened, the NUWSS declared that they would abandon active campaigning and devote themselves instead to the war effort. Sending out instructions to groups all over the country, Millicent Fawcett urged women to postpone their bid for the vote and instead defend the nation. 'Let us prove ourselves worthy of citizenship,' she proclaimed, 'whether our claim be recognised or not.'[7] Similarly, the militant suffragettes renounced violent action and committed themselves to supporting Britain. Prematurely rejoicing that suffrage 'battles are practically over', the militant Emmeline Pankhurst sought to convert her followers into patriotic soldiers by announcing that 'directly the threat of foreign war descended on our nation we declared a complete truce.'[8]

The straightforward version of events describes how women gradually and successfully overturned initial hostility to their involvement in the War. Benefiting from networks that had been built up over decades, suffrage campaigners were ideally placed to respond quickly by raising money and recruiting volunteers. As the male workforce diminished, and industry struggled to meet increased demand, women took over vital tasks necessary for supplying the country's technological needs. 'Who works, fights', declared Lloyd George in gender-neutral language, and the government publication *Women's War Work* listed ninety-six trades and 1,701 jobs.[9] Before long, women were rejoicing that public opinion had been transformed—'woman's place', declared a former suffragette jubilantly, 'is no longer the Home. It is the battlefield, the farm, the factory, the shop.'[10] The writer and journalist Rebecca West had opposed suffragette violence, but suggested she felt that it had given previously downtrodden campaigners

> a courage which now they use in the service of their country…[O]ne doubts that women would have gone into the dangerous high explosive factories, the engineering shops and the fields, and worked with quite such fidelity and enthusiasm if it had not been so vigorously affirmed by the suffragists in the last few years that women ought to be independent and courageous and capable.[11]

Unsurprisingly, this gratifying narrative skims over a complex network of conflicting interests. For one thing, neither women nor men were united by their gender. Traditionally, warfare was for men, and peace was for women. The staunchly pacifist Virginia Woolf later articulated this view to her nephew:

> if you insist on fighting to protect me, or 'our' country, let it be understood, soberly and rationally between us, that you are fighting to gratify a sex instinct which I cannot share; to procure benefits which I have not shared and probably will not share.…As a woman, I have no country…as a woman, my country is the whole world.[12]

A British soldier put it more bluntly, commenting that suffragettes should be shot, and continuing 'I've been afraid that if we let women vote, they might all vote against war.'[13] But in reality, there was no such neat division: there were men who refused to fight, and women who rose to the challenge of warfare. One suffragist expressed her contempt of female pacifists, exclaiming to her diary: 'How can English Women at present go and prattle with German Women of peace...? [W]hen thousands are laying down their lives for women to talk like that is to my mind showing a tremendous lack of nationalism.'[14]

Even the most patriotic of soldiers could feel as if they had been released from industrial exploitation only to become human cogs in the giant machine of warfare. Skilled workers—builders, miners, engineers—were often forced to accept a reduction in their pay packets. Would-be volunteers who had been rejected as unfit suffered public humiliation when over-enthusiastic suffragists handed them white feathers in tactless bids to shame the hesitant into action.[15] Returning invalids who had fallen sick after enduring months in the trenches attracted little sympathy compared with those whose gold stripe indicated they had been injured, whatever the cause.[16]

Class distinctions also came into play. In a misguided attempt to be patriotic, upper-class women stopped buying luxury goods—thus consigning to unemployed poverty the workers who had been producing them. As an additional self-sacrifice, privileged householders decided to economize by dismissing servants—thus swelling still further the numbers out of work. Some ladies of leisure busied themselves by retraining former maids and cleaners to make toys, including trenches looking 'for all the world as if they have been rained on and have become caked with the jolliest mud possible'—hardly suitable diversion for a child whose father was missing in action.[17]

As a further complication, it could suit the government to collude with women's organizations. In July 1915, Emmeline Pankhurst led a massive and much-publicized 'Right to Serve' procession down the Strand to the Ministry of Munitions, where her troops of former activists were warmly welcomed. This was no spontaneous protest, but a carefully staged media event proposed by Lloyd George and financially subsidized by the government.[18] Conversely, a few months earlier, the Admiralty had deliberately duped suffragists about an international women's Peace conference being planned in the Hague. When ferry services across the Channel were suspended for ten days, the official reason given was preventing women from attending the conference—but in actuality, this was part of an elaborate intelligence scheme to fool the Germans.[19] From the government's point of view, the more energy women put into arguing amongst themselves, the less forceful would be their claim to the vote.

Women Join the Scientific Workforce

Rapidly taking on the role of a patriotic employment agency, the NUWSS set about finding work for women who had lost their jobs. In London's Victoria Street, they established a Women's Service Bureau, hiring a large staff of interviewers who placed many thousands of women a year into a wide range of paid positions. Their most pressing immediate task was to care for the quarter of a million refugees from Belgium, a country often depicted as a young female victim raped by the German intruder. In addition to such traditional relief work, they also trained women to carry out scientific and technical tasks conventionally reserved for men—chemical manufacture, car maintenance, aeroplane production, engineering. For example, Mary Lowndes—the stained-glass artist who designed suffrage promotional material—founded a free training school for female oxyacetylene welders, who were taught professionally by experts from the Amalgamated Society of Engineers.[20]

By the time the War ended, the number of employed women had increased by well over a million—from around a quarter to a third of the workforce.[21] That sounds rather vague, but exact figures are impossible to ascertain because even official statistics were often massaged to fit a particular argument. Whereas the Ministry of Munitions wanted to boast about the success of its recruitment campaign for boosting productivity, other government departments tried to down-play the number of female workers because of worries about equal pay campaigns. Whatever the precise data might have been, there was no uniform growth: all the tables show striking disparities between occupations. The two largest industries employing women remained the same as before—textiles and clothing. But they continued to shrink during the War as the decline in the fashion industry outpaced the demand for uniforms. In contrast, by the end of the War, the civil service included over 200,000 women, more than one hundred times the initial figure. And whereas the large armaments factory at Woolwich had previously employed around 10,000 men, its workforce swelled to reached 75,000, over a third of them women.[22]

About 90 per cent of the women who entered engineering and related trades were employed in areas not previously regarded as suitable for women. The biggest single domain was munitions, an umbrella term covering two main products: cordite, a low explosive that had already replaced gunpowder as a propellant; and TNT, a high explosive previously made in Germany that was needed to fill shells. Located near raw materials, the factories were concentrated in two bands, one stretching across northern England between Manchester and Leeds, and the other near London along the Thames. In contrast with Germany, many factories were new and run by the state, often constructed in neo-Georgian styles to reflect national pride in defending the nation. But others were explicitly temporary: the women knew that they would no longer be employed after the War. And whereas the French production plants were staffed by wounded and convalescent servicemen, in Britain it was women who carried out this dangerous and tedious scientific work.[23]

To overcome employer resistance to scientific women, government propaganda emphasized traditionally feminine virtues, quoting a factory manager's comment that 'Girls are more diligent on work within their capacity than boys—they are keen to do as much as possible, are more easily trained.'[24] But as the Daily Chronicle put it, a munitions factory was 'always full of people and yet is the home of nothing except death'.[25] The complex at Gretna, which straddled the border between Scotland and England, was the largest industrial site in the entire world, protected by barbed-wire fences and sentry patrols— possibly to retain secrecy but also preventing munitionettes from leaving and local men from entering. Hastily constructed by thousands of Irish labourers, it stretched across a vast area nine miles long and two miles wide so that buildings could be isolated in case of explosions. Gretna was almost entirely staffed by women, who worked in twelve-hour stints to keep the machinery constantly running and maximize their earnings. They lived in groups of huts linked by raised wooden gangways, their beds rationed out in shifts to save on accommodation costs.

The journalist Rebecca West made the process of manufacturing cordite sound like kitchen cookery, reporting that the women resembled millers in their white waterproof dresses and hats, while the 'brown cordite paste itself looks as if it might turn into very pleasant honey-cakes'. She described these uniformed chefs as 'pretty young girls clad in a Red-Riding-Hood fancy dress of khaki and scarlet'. But she also emphasized the sacrifices they were making while their men confronted danger at the front. When two huts were gutted after an explosion, 'the girls had to walk out through the flames. In spite of their chemically fire-proofed clothes, one girl lost a hand.'[26]

Wartime women took over men's jobs in factories, and they also stepped out onto the sports pitch. As part of a puritanical attack on drink, prostitution, and entertainment, professional football was suspended at the end of the 1914–15 season, but the Munitionettes' Cup attracted thousands of spectators to watch female football teams representing different factories. Ladies' football had been introduced towards the end of the nineteenth century, and initially aroused immense outrage. After a game in 1895, players were castigated for appearing in public wearing coloured knickerbockers, and the *British Medical Journal* warned primly that doctors could 'in no way sanction reckless exposure to violence, of organs which the common experience of women had led them in every way to protect'. The War helped to make female football respectable.[27]

That sounds like a terrific success story for women—which was exactly how many people chose to present it. 'Oh! This War! How it is tearing down walls and barriers, and battering in fast shut doors,' enthused a female journalist keen to get across a positive message about opportunities.[28] In the summer of 1916, speaking in his role as

Minister of Munitions, Lloyd George told the House of Commons that 'it is not too much to say that our Armies have been saved and victory assured by the women in the munition factories where they helped to produce aeroplanes, howitzer bombs, shrapnel bullets, shells, machine tools, mines, and have taken part in shipbuilding.'[29] Other propagandists transported heroic accolades from the Western Front to the Woolwich factories, proclaiming that 'the women of Great Britain are as ready as their men folk to sacrifice comfort and personal convenience to the demands of a great cause.'[30] Optimism ran high: 'A man drops dead in the trenches,' enthused the American journalist Mabel Daggett; 'Back home another woman who had been temporarily enrolled in the ranks of industry steps forward, enlisted for life in the army of labour.'[31] She spoke too soon: once the War was over, most of them would lose their jobs to men.

Recruitment posters disguised employment realities by showing cheerful, energetic young women resourcefully saving the nation: one ploughs a sun-drenched field, another shrugs on a luxurious coat as she waves to a passing soldier, a third jubilantly stretches her arms open to the sky, reflecting the shape of the aeroplanes flying overhead. Left behind and in a now predominantly female environment, women wrote and performed plays designed to keep up their spirits. Referring to the vast numbers of women employed in munitions factories, Gladys Davidson reminded men in the audience to recognize the home war effort—the production line that kept the front line going:

> You're heroes, ev'ry one!
> But if it weren't for Vulcan and the Girls, where would you all be, I wonder?
> Don't forget that we make the bullets for you to fire;
> And when you come back home again, you may find us wearing medals, too, for sticking to our job![32]

Photography was banned at the front, so newspapers sought to attract readers with appealing pictures of feisty women labouring to support the War effort. Their hair wrapped up in headscarves, their bodies hampered by lengthy skirts, former domestic servants were captured by the camera frowning in concentration as they adjust precision equipment, trundle loaded wheelbarrows, delve beneath car bonnets, grapple with heavy machinery, and stack up large piles of shells. Government officers toured provincial factories with albums designed to convince reluctant employers that women could undertake skilled labour (and that yes, it was indeed worthwhile investing in separate lavatories). A century later, these photographs provide a rich resource for television programmes, books, and museums, which display the nostalgically grainy black-and-white images as if they were neutral depictions rather than instruments of propaganda.[33]

Class differences have often been sanitized out of war recollections, but they permeated wartime experiences. In many factories, Lady Bountifuls took on the task of unpaid welfare supervisors to protect the labourers, who not surprisingly resented 'voluntary workers, they are such busybodies. They always sum up the girls into the nice girls and the rude girls and the nice girls get attention and the rude girls get nothing.'[34] Badgered by his daughter, Thomas Pullinger—an industrialist in a Scottish village—enticed middle-class recruits from all over the country by promising to train them for future careers in engineering. 'Shell making is . . . a cul-de-sac', his advertising material correctly commented; in contrast, he promised permanent positions 'leading not only to good emoluments but high positions in the industry'. Opening in 1917, his factory produced aeroplanes, the glamorous end of wartime manufacturing, and he employed sixty college-trained women, who also received professional on-site instruction. Wearing suffrage (non-militant) badges of red, white, and green, these fast-track apprentices initially took great pride in their skills. But Pullinger's experiment soon foundered. As genteel young women, they resented their low pay, were bored by the repetitive tasks, and felt isolated in a rural backwater. Before long, the lectures ground to a halt, and local women replaced the ungrateful recruits.[35]

The euphemism 'unoccupied' denoted all those privileged women who had never needed to work but were now eager to volunteer. Many of them effectively played at becoming patriotic labourers. The Prime Minister's daughter-in-law, Cynthia Asquith, enthused about the two hours she had spent making respirators: 'It is such fun being a factory girl,' she exclaimed; 'how exciting it must be to do piece-work for money.' But realism soon returned. She continued 'I must say I was very glad I hadn't got to do a twelve-hour day—it is quite tiring.'[36] One self-sufficiency manual written for the new 'bachelor girls' spelt out in great detail basic information about planning a budget, removing stains, or cooking peas in a jam-jar. Almost a whole page was devoted to the tricky task of making a cup of tea, and over three to washing up. Possibly the section that got referred to most often was the one explaining how to pay a charwoman.[37] Even those not subjected to the ordeals of manual labour felt that they were hard done by: 'Not a parlourmaid to be had', lamented Virginia Woolf in 1917.[38]

Whatever their class origins, many women looked back nostalgic-ally at the War, often relying on conventional clichés to portray it as the best period of their lives. Released from the boredom of being a middle-class housewife, a mechanic in a London aeroplane factory found the work so stimulating that 'it was nothing to leap out of bed at 5.15 on a frosty morning…I almost danced down Queen's Road under the stars, at the prospect of the day's work.'[39] A Coventry ordnance employee demanded more training so that she could make Howitzers, later remembering it 'as a very happy time and well, every-thing was very happy, the atmosphere in work and the people you worked with'.[40]

Even if such reactions were reported honestly, how can we be sure that they were not rose-tinted memories by an unrepresentative few? Those ravaged by survivor guilt are often reluctant to recall suffering, and historians may be tempted to seek out positive mes-sages affirming female solidarity. The author of a prize-winning essay published in the magazine of a projectile factory expressed an ambivalence that was perhaps shared by many. Torn between conflicting views, she admitted

> the fact that I am using my life's energy to destroy human souls gets on my nerves. Yet on the other hand, I'm doing what I can to bring this horrible affair to an end. But once the War is over, never in creation will I do the same thing again.[41]

These women may have enjoyed the camaraderie of factory life, but they knew they were dispensing death to other mothers' sons.

Difficulties and Dangers

When war was declared in August 1914, many people immediately volunteered. But whereas young men like the physicist Henry Moseley and the poet Wilfred Owen achieved their dream of being sent into action, their female peers felt as if they were being treated like children. The Oxford undergraduate Vera Brittain volunteered to nurse in France, but although her student fiancé died at the front, she was summoned home to care for her ageing parents as if her war work were trivial compared with her duties as a daughter. Rather than being valued as active participants in affairs of the moment, young women were reserved as potential mothers for soldiers of the future. '[P]erhaps the most thrilling event of the term from the feminine point of view has been the appeal for women munitions-makers,' wrote a bitterly disillusioned Newcastle university student in 1916.

> After twenty-four distracted and heart-searching hours; after some of us had obtained permission to go and were feeling heroic: after some of us had decided to stay and were feeling 'desper'te mean'; and after some of us had defied our obdurate parents and guardians in the names of Patriotism and Legal Majority, and were feeling Shelleyesque, we were informed that our highest patriotic duty was to complete our education![42]

In protests against the horrors of chemical warfare, the sufferings of munitionettes were scarcely mentioned. Attention focused instead on poison gas that affected men. Wilfred Owen gave his famous war poem an ironic title—*Dulce et Decorum Est*, Latin for 'It is Sweet and

Honourable' and part of a quotation from the poet Horace about dying for one's country. This stanza is often reproduced:

> Gas! Gas! Quick, boys!—An ecstasy of fumbling,
> Fitting the clumsy helmets just in time;
> But someone still was yelling out and stumbling,
> And flound'ring like a man in fire or lime...
> Dim, through the misty panes and thick green light,
> As under a green sea, I saw him drowning.
> In all my dreams, before my helpless sight,
> He plunges at me, guttering, choking, drowning.[43]

Despite the awfulness of gas, military experts on both sides pointed out that, unlike bullets, it was not usually fatal and the effects were temporary. Less commemorated are the munitionettes who worked with hazardous chemicals that often permanently affected their health. In a factory attached to a small Welsh mining village, the ether used to make cordite was particularly damaging. 'It gives some headaches, hysteria, & sometimes fits,' a police woman reported; 'if they stay on in the cordite sheds, they become confirmed epileptics & have the fits even when not exposed to the fumes. Some of the girls have 12 fits or more one after the other.'[44] TNT was also pernicious. As well as causing nausea, headaches, and rashes, it made hair turn green and 'our skin was perfectly yellow, right down through the body, legs and toenails even, perfectly yellow'.[45] Nicknamed canaries, victims were relegated to separate canteens in the factories because 'everything they touched went yellow, chairs, tables, everything'; when they travelled by train, they found themselves shunted into separate carriages to avoid disturbing fellow passengers.[46] Around 200 of these women died, either from chronic poisoning or in accidental explosions; one survivor, Lilian Miles, recalled that 'you'd wash and wash and it didn't make no difference. It didn't come off. Your whole body was yellow.'[47]

Even other women regarded them with disgust, a very different emotion from the sympathetic admiration reserved for afflicted men. Waiting for a train, the pacifist Caroline Playne remembered seeing a sight foreign to anything civilization has ever exhibited before...a couple of hundred de-humanized females, Amazonian beings bereft of reason or feeling, judging by the set of their faces, bereft of all charm of appearance, clothed anyhow, skin stained a yellow-brown even to the roots of their dishevelled hair....Were these really women?[48]

These chemical casualties resented being stared at in public, but being spurned by their own families probably felt worse—and then there was the worry that their internal organs had also been stained the sickly green tint. Quick to take advantage of the munitionettes' concerns, advertisers unrealistically promised that Oatine face cream or Mrs Pomeroy's Skin Food would keep their skin soft and prevent 'the munition complexion'.[49]

As news of the toxic working conditions spread, the Ministry of Munitions tried to maintain recruitment levels by appointing factory doctors. Their role was not so much to protect the workforce, but more to reassure the munitionettes by (supposedly) checking their health. Whereas the 'canaries' and their families agonized over their abnormally yellow hair and skin, the doctors apparently accepted the weird colouring as a natural consequence of handling TNT. Instructed to ignore the women's own reports on how they felt, these medical inspectors were encouraged to identify those who were experiencing merely minor effects, so that money could be saved by sending them back to work.

Sufferers were only deemed genuinely ill when their symptoms had worsened sufficiently to match this diagnostic template: 'a pale face lacking in expression and like anaemia but peculiar in itself, lips that can hardly be described as cyanosed but of an ashen blue colour, similar gums, and perhaps a faint trace of yellow on the conjunctivae'.[50] By that stage, the prognosis was poor and even those who survived never fully recovered their health. Even so, in her post-War report for the War Cabinet, a female doctor seems to have been as concerned about the welfare of the nation as of the women: 'The stunted physique and poor health of many of our industrial workers suggests there may also be the graver and more fundamental injury to the germ plasm itself and so to the future of the race.'[51]

Adding to the hazards of toxic chemicals and industrial machinery, the risk of explosions was high. Women were carefully sheltered from atrocities at the front, but they might witness appalling scenes of slaughter in their own neighbourhoods. Laconic diary entries implied that heroic women could take all this in their stride: 'The shed was wrecked & the machine blown to chips & the girls shot out on to the bank opposite. One found that her shoes were alight so calmly kicked them off & both trotted up to the canteen. Meanwhile agitated firemen were hunting for their dismembered remains.'[52] The government tried to play down the number of reported casualties, but they could hardly conceal one of the worst, which occurred in the East End of London. The light from exploding chemicals was visible for miles around, and the shoddy houses in nearby streets collapsed into ruins. Perhaps most frightening of all, flames licked across the Thames towards the nearby Woolwich Arsenal, blowing up a gasometer.

Women's work was boring, yet dangerous. 'Small bits of brass seemed to be a target for my eyes,' recalled Edith Airey; 'I was for ever at the first aid post having bits extracted.'[53] Almost 90 per cent of industrial chemists were women, and they were assigned tedious, repetitive tasks designed to resemble following a recipe—the same one, day after day, month after month. The war artist Christopher Nevinson showed female acetylene welders sitting along a production line like identical robots, their eyes protected by large goggles and their hair gathered under a cap (see Figure 5.1). In 1916, Sheffield University introduced a 'special one-month Intensive Course' to train women for 'routine repetition work'. Over the next two years, ninety-six women completed it, and unusually, many of them kept their jobs after the War as 'an undoubted success in this branch of industry'.[54]

Welding an aeroplane might sound exciting, but gluing fabric on the wings was monotonous and entailed painting with poison. Official recruitment material minimized the risks. Glossing over the hazards of working with metals, a Ministry of Labour leaflet reported that 'The average woman takes to welding as easily as she takes to knitting once she has overcome any initial nervousness due to sparks.'[55] Compare

Figure 5.1. Christopher Nevinson, *The Acetylene Welder*, 1917, Lithograph.

that bland assurance with this report by a policewoman sent to a Chester factory:

> The particles of acid land on your face & make you nearly mad, like pins & needles only much more so; & they land on your clothes & make brown specks all over them, & they rot your handkerchiefs & get up your nose & down your throat & into your eyes, so that you are blind and speechless by the time your hour is up & you make your escape.[56]

Protected from danger, privileged writers patronizingly praised those who preferred to suffer rather than be put out of work: a 'working woman recently lost the first finger and thumb of her left hand, owing to a loaded gaigne [*sic*] jamming in the press. After an absence of six weeks she returned to work....'[57]

Unequal Treatment

After it became apparent that the fighting was going to last for more than a few months, opposition to female employment started to melt, although newspaper articles fanned fears of role reversal: 'the danger is that at the end of the war the masterful woman will be in control of the feminine one to the detriment of the race.'[58] Soldiers serving abroad worried about the masculinization of the women they had left behind. 'I think it disgusting the way women are going on at home,' George Wilby wrote to his fiancée; 'When us chaps come home we shan't know the women from the men.' He was also worried about her moral decline: 'I'm so afraid if you go into a factory or any other rough work it will take all the womanly qualities out of you, Dear, and leave you coarse and unlovable like most other girls.'[59] When he discovered she had disobeyed his pleas by making shells, he accused her of betraying her destiny as a mother. Although he was risking his life, he promised that they would 'make up for the damage *you* [his emphasis] have done'.

Wilby's girlfriend probably suffered similar taunts from the older and medically disqualified men left behind to run the factories, who resented what they saw as an invasion of their masculine territory. Resistance to female labour was fuelled by trade union officials concerned to protect the status and pay structure of their men, so that debates about differential pay schemes continued right through the War and for decades afterwards.[60] 'Engineering mankind is possessed of the unshakeable opinion that no woman can have the mechanical sense,' complained a resourceful female recruit; 'Any swerving from the path prepared for us by our males arouses the most scathing contempt.' She transferred into the workplace the traditional womanly skills of managing an obdurate husband or father. 'As long as we do exactly what we are told and do not attempt to use our brains we give entire satisfaction and are treated as nice good children,' she wrote, going on to explain that men 'have to be managed with the utmost caution lest they should actually imagine it was being suggested that women could do their work equally well, given equal conditions of training—at least where muscle is not the driving force'.[61]

Once it became clear that women were running Britain's factories, they became subjects of scientific study. To make these women efficient—to get the best value from them—they needed to be healthy. Chemists formulated well-balanced menus, doctors prescribed better ventilation and bathrooms (including carefully angled showers to prevent long hair from being soaked), and factory inspectors carried out research showing that—unsurprisingly—women worked better if they were given rest breaks and cloakrooms to change out of wet clothes.

The temporary influx of women permanently changed the industrial landscape. To compensate for women's lesser physical strength, factories introduced mechanical equipment. One shell factory manager minimized female initiative by explaining that 'We put the brains into the machine before the women begin', but after the War this equipment also improved male efficiency.[62] Female novices were frustrated by men's automatic assumption that a mere woman could not suggest any technical improvement:

> If one of us ask humbly why such and such an alteration is not made to prevent this or that drawback to a machine she is told with a superior smile that a man has worked the machine before her for years and therefore that if there were any improvement possible it would have been made.[63]

Refusing to be cowed by such dodgy logic, the female invaders actively sought professional training and managed to increase productivity substantially. Hard-working and enthusiastic, they implemented better ways of carrying out existing tasks as well as developing new chemicals, weapons, and machine-parts.

Then as ever, women's appearance was a matter of great concern, and there were numerous complaints about women modifying their uniforms by shortening their skirts or lowering the necks of their blouses. A new disease was invented—khakitis, a pejorative term for fashion-conscious young women who dated soldiers in uniform and were accused of behaving like prostitutes and of spreading sexual diseases. Manufacturers were quick to cash in. A fashionable corsetmaker marketed horribly restrictive underwear that was guaranteed to provide the latest 'Military Curve', while advertisements for Vinolia face cream featured young women wielding spanners or operating lathes: 'Beauty on Duty has a Duty to Beauty', ran the slogan. By the time the War ended, working women had forced dress designers to modify their styles. The London stores were now selling shorter skirts, loosely fitted smocks and jackets, and sturdy waterproof boots. Whalebone corsets were out, broad-shouldered trench coats were in.[64]

What might sound like compliments often had scornful under-tones, so that however they dressed, women fell prey to mocking remarks. Nurses on motorbikes wore helmets and breeches instead of starched headdresses and regulation skirts, but they were saluted ambiguously: 'Wonderful little Valkyries in knickerbockers, I take off my hat to you.'[65] When female welders adopted protective clothing, they were accused of retaining their feminine wiles: 'Overalled, leather-aproned, capped and goggled—displaying nevertheless the woman's genius for making herself attractive in whatsoever working guise.'[66] Historians have perpetuated such condescension—this is from a book published in 1967: 'The new pride in personal appearance, especially among a class of women who had formerly become ill-kempt sluts by their mid-twenties, if indeed they had ever been anything else, was one of the pleasantest by-products of the new female self-confidence remarked upon by the Chief Factory Inspector.'[67]

Women were also often critical of female workers who cropped their hair, even though that was one way of avoiding lice and accidents. Not only did they look like men, ran the complaints, they 'assumed mannish attitudes, stood with legs apart...if they cannot become nurses or ward maids in hospitals, let them put on sunbonnets and print frocks and go and make hay or pick fruit or make jam.'[68] Trousers were another practical choice, but many women preferred to follow convention and avoid ridicule by struggling to work in cumbersome skirts totally unsuitable for heavy labour. At Woolwich Arsenal, a female graduate was impressed by the workers' attempts to challenge their drab environment, describing how they flouted regulations by wearing flowers or swapping their official black shoelaces for ones in brilliant colours.[69]

Money was another contentious topic.[70] Both during and after the War, the government and other industrial employers concocted devious reasons for maintaining large pay differentials. Although official documents dressed up their arguments in management-speak, they expressed essentially the same point of view as male factory workers. When one lathe operator dared to complain, her foreman justified paying her less than her male neighbours who were producing fewer shells per hour. 'A man', he explained, 'has got a wife and children. He needs more money than a girl, and women are cheaper than men because their wants are fewer. For instance, they don't require tobacco; and tea and toast is cheaper than beer and beefsteaks.'[71] Unsurprisingly, factory women trying to hold their families together saw it differently: 'Old man sitting on the Arsenal wall, Just like his dad doing sod all.'[72] Whatever the realities of their smoking and eating needs, many wives were bringing up orphaned children or caring for elderly relatives. In any case, as the cost of living escalated during the War, a husband's wages were no longer enough to support a large family.

Whether based in laboratories or factories, in universities or heavy industries, women were paid substantially less than men, and even those with professional qualifications often acted as unpaid volunteers. Initially, employers argued that because female novices were unskilled and physically weaker than men, they should be paid two-thirds of the standard wage. In principle, this claim should have become discredited once female apprentices had been trained to carry out the same work as the men, especially after machinery was installed to take over heavy lifting. In practice, a common yet circular no-win argument came into play: any jobs suitable for a woman must demand less expertise and hence deserve lower wages.

According to a trades union report, 'the meanest episode in the sordid story' was when the government backed employers in their bid to defeat the claims for equal pay being made by the Society of Women Welders. Before women were recruited, welding had been classed as skilled labour, but the first female trainees were asked to accept half the hourly rate paid to men. With union backing, the Society managed to get the rates increased slightly, but disputes rumbled on throughout the War and beyond. When work was short, men were worried about being undercut by women, who in their turn were angry about being paid less. As a letter from the Society explained, the opposition was slippery: 'we claimed that the women were doing skilled men's work and should receive the same wages; the employers were unable to claim that women were not doing the same work, and therefore claimed that it was not a skilled job.' That tactic was widely deployed: as soon as a woman was trained to carry out a particular task, it was reclassified as unskilled and set on a lower pay scale. After negotiations were interrupted by the Armistice, the government ruled that in the interests of keeping down wages, it was more important to maintain continuity than to establish an unsettling new principle—that women should receive equal pay for equal work.[73]

In contrast, young single women recently released from domestic servitude were enjoying for the first time what it felt like to have spare cash at the end of the week. Often encouraged by their parents, teenagers entered the factories and the laboratories to acquire new skills—and also new clothes. Whatever they wore, they were criticized. Some newspaper articles stoked rumours that factory labourers were buying expensive jewellery and replacing their tatty shawls with sumptuous furs, while others fumed that 'a great many of our young munitioners are what Lady Macbeth prayed to be, unsexed from the crown of their bonneted heads to the toe of their masculine overalls....They are pampered. The masculine dress has a psychological effect on the susceptible.'[74] Resentment seethed among puritanical reformers, and the League of Decency and Honour recommended restricting entertainments as well as banning silken underwear or perfumed soaps. Retaliating, one local paper sharply reprimanded these privileged ladies, commenting that they 'ought to mind their own business, and instead of joining the "League of Honour" should join the "League of Hard Work", and then they would not have so much time to bother about other people's affairs'.[75]

Writing in the *Daily Express*, a particularly articulate munitions worker protested that even if she patriotically invested in war certificates, she would be accused of earning too much money for her station in life. 'It is the old autocratic spirit,' she proclaimed, 'struggling in its death throes to make a last endeavour to assert itself. The same spirit which my efforts in war work are supposed to be helping to crush.'[76] If she survived, the author of this ringing manifesto presumably became still more angry and disillusioned after the War. Patriotism may have united the nation for four years, but although attitudes towards women had altered, true equality—whether of class or of gender—remained an unfulfilled dream.

6

RAY COSTELLOE/STRACHEY

The Life of a Mathematical Suffragist

The Queen is most anxious to enlist every one who can speak or write
to join in checking this mad, wicked folly of 'Woman's Rights', with all
its attendant horrors, on which her poor feeble sex is bent, forgetting
every sense of womanly feeling and propriety.

Queen Victoria, letter to Theodore Martin, 1870

A *Room of One's Own* is among the most famous feminist texts ever written, but during the First World War few people had heard of its author. By the time Woolf's book appeared in 1929, Ray Strachey had already published six of her own, including *The Cause* (1928), celebrated for decades as the standard history of the early women's movement. A leading campaigner in the suffrage movement, she had stood for Parliament three times, and was renowned for making sure that change actually happened. As a principal of Somerville College remarked, Strachey was 'someone who takes up lost causes and then they cease to be lost'.[1]

Virginia Woolf, Duncan Grant, Vanessa Bell, Lytton Strachey—the Bloomsbury Set casts such a gilded glow over the early decades of the twentieth century that it is easy to forget their contemporaries. The posthumous glamour of this apparently charmed circle outshines unpromising beginnings. In 1909, three years before she became Mrs Woolf, Virginia Stephen invited the Cambridge mathematics graduate Ray Costelloe to visit her in London. Both in their twenties, both of them unmarried, the young women liked each other, but

Costelloe—who stemmed from a family of American Quakers—had reservations about her hostess's friends, describing them as 'a queer self-absorbed, fantastic set of people!'.[2] She commented in bemusement that after an evening's idle chatter round the fire, they 'somehow fell to making noises at the dog, & this august awe inspiring company might have been seen leaping from chair to chair uttering wild growls amid shrieks of laughter'.[3] Within a few weeks, she had discovered 'They don't seem to do very much, these people, but they talk with surprising frankness. One of their greatest joys seems to be to tear their friends limb from limb.'[4]

The two women could hardly have been more different. Keen on engineering and sport, Costelloe was a happy optimistic person with little interest in clothes yet careless about overspending or running into debt. Costelloe found Stephen to be 'a very charming person. She spends all her time reading & writing, does no suffrage work',[5] while Stephen assessed Costelloe as 'a good satisfactory creature, as downright as a poker'.[6] A couple of years later, still both single, they spent a gossipy evening together talking about men and love. After some gentle probing, Stephen discovered to her relief that there was no affair brewing between Costelloe and her brother Adrian—'perhaps one would find her a little heavy' as a sister-in-law, she wrote to her sister Vanessa.[7]

By the end of the War, their relationship was very different. Ray Strachey enjoyed a national reputation as one of the pioneering campaigners who had persuaded suffragists to fight for Britain as well as the vote. In contrast, Virginia Woolf was little known outside her own set of friends. Ray 'makes me feel like a faint autumnal mist,' she wrote plaintively; 'she's so effective, and thinks me such a goose.'[8] On the other hand, they were now permanently tied together by marriage.

The Bloomsbury industry has neglected the two Costelloe sisters—usually referred to by their married names of Ray Strachey and Karin Stephen—who lay at the centre of the famous set and formed a crucial strand in the intricate web of sexual relationships knitting it together. It was thanks to them that the two leading families, the Stephens and the Stracheys, were formally connected in 1914, when Adrian Stephen became the husband not of Ray, as Virginia had feared, but of her younger sister, Karin. Ray herself had already married Lytton Strachey's younger brother Oliver, and the couple's main marital home was literally in central Bloomsbury—in Gordon Square, whose other residents included Lady Strachey, the Stephens, Clive Bell, and John Maynard Keynes. Although the Costelloe sisters' relationship remained antagonistic, their husbands became so close that they founded Soho's 1917 Club to honour the Russian Revolution and celebrate the fall of the tsarist dynasty.[9] Collectively, their lives seem to confirm Dorothy Parker's apocryphal aphorism that the Bloomsbury set lived in squares, painted in circles, and loved in triangles.

Becoming a Strachey

When Ray Costelloe met her husband-to-be Oliver, the youngest of the eleven Strachey siblings, she had already chosen the family she wanted to join (since there were so many Costelloes and Stracheys, I will use first names).[10] For her, it was love at first sight with all of them rather than with Oliver in particular. Invited to their Hampstead home by her Newnham friend Ellie Rendel, who was Lytton Strachey's niece, she was immediately captivated by their casual behaviour—the eligible bachelors who shunned female visitors, the seventy-year-old matriarch who propped her book up against her drinking glass at dinner (a habit Ray acquired and passed on to her children). All very different from life at home![11] She felt oppressed by her own family, who mockingly dismissed her bid for privacy by nicknaming her 'The Oyster'. 'It is only the personal things they are after,' Ray told her diary when a teenager; 'That horrible way in which people ask you questions and expect you to pour out the secrets of your heart—and then if you do in any measure, they patronize you and torture you ever after.'[12] In her novel about the American Civil War, she may well have been thinking of her own childhood when she created a heroine who

'stumbled through life, awkward, vague and incompetent, and...
developed absent-mindedness to protect herself from the world'.[13]

It was through the Strachey clan that Ray became actively involved
in campaigning for suffrage. She had already been strongly influenced
by her own family's fiercely independent women, who were early
supporters of the movement. In particular, she adored her grandmother Hannah Whitall Smith, a thin Pennsylvanian Quaker of ram-
rod posture who favoured sombre clothes and wrote propaganda for
the World's Women's Christian Temperance Union—watchwords
'*Agitate! Organize!*'[14] After Ray's mother Mary left her first husband
and her two daughters in London to live in Florence, it was essentially
Hannah who brought Ray up from the age of four. A few months
before she went up to Cambridge, Ray told grandmother Hannah that
she had learnt three survival lessons from her: 'you *must* have plenty of
sleep, and you *must* take care of your health, and you *never* must get
offended.'[15]

Clearly the source of power in the family, Hannah indulged her
granddaughters but insisted she knew better than their teachers and
doctors, whom she openly mocked. She often annoyed men—notably
her son-in-law Bertrand Russell, the husband of Ray's aunt Alys.
Hannah once told Alys she would be attending classes by Mary
Boole, wife of the famous mathematician, to learn about Russell's
philosophy; 'Perhaps—who knows—I shall understand him in some
small degree afterwards,' she wrote drily.[16] As a young adult, Ray
spent many hours discussing politics and suffrage with 'Uncle Bertie',
who in 1910 stood (unsuccessfully) in a by-election as a candidate
supporting suffrage.[17] Even so, she worried about never being quite
sure what he really thought—perhaps just as well, since he judged her
to be like his wife: 'kind, hardworking, insincere and treacherous'.[18]

Ray's uncle was the eminent literary critic Logan Pearsall Smith,
and she was named after another aunt who had died. To avoid
having further children—or perhaps already disillusioned with their
marriage—her parents reportedly abstained from sexual intercourse
for months, adamantly refusing to consider any form of birth control,

then widely regarded as immoral.[19] Although Karin was born a couple of years later, Ray remained the favourite daughter, a pressure that she loathed. She also resented the frequent attempts to make her more conventionally feminine by telling her to dress more elegantly and move more gracefully. She loved reading, but her main intellectual delight was mathematics. 'It's like a fairy land,' she recorded privately,

> it gives me a queer feeling of insignificance.... [I]t is just the great solid & infinite Truth standing calmly still to be realized. It is not affected by my mood & powers, and continues sublimely on, self sufficient. I wonder no one has made a god of Mathematics.[20]

Her other passion was sport—cricket, swimming, hockey—and she became head girl at her Kensington school. Hannah was delighted when Ray joined a Fancy Dress Party committee, fondly imagining that she was 'starting off on her independent committee career in this way. How I would love to attend one of their meetings. I fancy Ray's tongue is unloosed at such times.'[21] The proud, controlling grandmother had two valuable tips to pass on from her own experience of a lifetime's work in philanthropic and temperance organizations: '(1) Be perfectly willing to give your best advice; (2) Be perfectly content not to have it followed.'[22] Ray benefited from that guidance when she was later immersed in suffragist arguments. 'I often thank Heaven for my equable temper,' she wrote to her mother, 'for some of the[m] are extraordinarily trying.'[23] With experience, she acquired another tactic for dealing with recalcitrant committees: 'the only thing is to wait till the old birds die off, or actually collapse.'[24]

Ray's mother Mary was also a great organizer, and the two often clashed. Mary had divorced Ray's father, the politician Frank Costelloe, to live with and later marry Bernard Berenson, a distinguished art historian who owned Villa I Tatti on the edge of Florence. During the sixty-odd years he lived there, Berenson converted the original farmhouse into an informal study centre for Italian Renaissance art. When Ray went to stay with 'Uncle Bernhard' for four months between school and Cambridge (her grandmother smuggled a back straightener into the suitcase), she was appalled by the young man her mother had singled out for her. Mary tried to rescue the situation by arranging a cultural trip to Venice, but Ray refused to look at any art. She was, however, thrilled at the opportunity of learning not only how to drive, but also how to crawl under the car and fix the engine, an experience she drew on for her first novel, *The World at Eighteen*.

By then, her mother was already disappointed that Ray had chosen to turn her back on fashionable literati and go to Newnham, but as Hannah told her, 'You can no more Strasburg Goose these young things into liking your social things than you can into liking your art things.'[25] Planning to study mathematics, Ray took extra lessons during the Christmas holiday, protesting to her sceptical aunt 'the more Mathematics I learn, the more I wonder that anyone can call them dull or useless. They seem to me to be the most exciting & beautiful ideas, exquisitely expressed.'[26]

When Ray went to Newnham in autumn 1905, it was Hannah who took her shopping, supervised the packing, and sent up supplies of chocolates, cakes, and jam. Two years later at Cambridge, Ray was still enthusiastic, reassuring her aunt that mathematics 'get more & more wonderful as I go on, and I am more than content to have chosen to study them'.[27] As finals approached, she confessed her concerns: 'I am very bad at my work unfortunately—but I am still convinced that it is a most fascinating subject. The more I do the more fascinating it seems. And I am working hard now, so I hope that I shall not fail.'[28] She scraped through—finishing eightieth out of eighty-six—still 'very much interested in the mechanical branches of Mathematics, & should like to qualify as an electrical engineer ultimately'.[29]

Ray's undergraduate career was unusual but not unique. When she went up to Newnham, around 20 per cent of Cambridge mathematics students were female.[30] Her examination results probably suffered because on top of the cricket and the hockey, she had been swept into suffrage activities by her close friend from school, Ellie Rendel. This was now her 'chief occupation', Ray enthused; 'We have joined with Girton college, & have over 250 members, and we have been sending circulars to all the past students of Newnham and Girton.'[31] The year after she graduated, she 'plunged' an aunt into 'Suffrage entertainments...I mean the Bernard Shaw play'. Presumably she meant *Mrs Warren's Profession*, initially banned because of its explicit references to prostitution. If so, she must have felt as if she were watching herself. Vivie Warren, the central character, was also a mathematics student at Newnham, similarly bored by art galleries, and aware that being at Cambridge 'means grind, grind, grind for six to eight hours a day at mathematics, and nothing but mathematics.... Outside mathematics, lawn-tennis, eating, sleeping, cycling, and walking, I'm a more ignorant barbarian than any woman could possibly be who hadn't gone in for the tripos.' At first, Shaw seems to be sympathizing with reactionary critics by poking fun at this self-styled 'perfectly splendid modern young lady'. But in a series of plot twists, Warren emerges as the play's heroine, the only one with sufficient strength of spirit to make hard moral choices. Taking a position as an actuary for a female barrister, she enjoys hard work, earns her own living, and relaxes over a detective novel with a glass of whisky and a cigar.[32] That was the type of independent life Ray aspired to attain— and she succeeded.

In February 1907, Ray joined the 3,000 women who tramped through London on the NUWSS 'Mud March', which was organized by her future sister-in-law, Ellie's aunt Philippa (Pippa) Strachey. Fifteen years older than Ray, Philippa became a major role model and collaborator, and it was under her influence that Ray became permanently committed to suffrage activity. Only the following year, she was back in London for another procession, this time carrying a Joan of Arc banner and accompanied by almost 400 members of Newnham sporting blue shoulder ribbons to symbolize that they were not allowed to graduate formally. Her mother strongly disapproved, reprimanding her in the Quaker idiom the family used among themselves: 'I am appalled to think of thy going in for politics in a "serious way"...I am sure it will bore thee to death.'[33]

After graduating, Ray decided that she and Ellie should visit her American aunt, who was President of Bryn Mawr, a women's college in Pennsylvania founded by the Quakers in 1885. Within a few weeks, she was immersed in suffrage activity, travelling around the USA and having a marvellous time, despite her growing recognition that conversations with strangers can be repetitive and boring. 'The fashionable suffrage meeting in New York was very aggravating,' she griped, 'because the smart ladies were so remarkably silly, and yet thought they knew all about the subject.'[34] Back in England, she resolved to give up suffrage work, but was soon sucked back in again.

From her mother's perspective, matters continued to deteriorate. That autumn, Ray enrolled in an electrical engineering class at Oxford, exulting in 'my lectures & labs—the dynamos & motors—roofs & bridges', the only woman in a class whose twenty men 'pulled twenty long faces'.[35] The following February she cut her hair short, and then shocked her sister by wearing a dirty blouse for a Bloomsbury party held in the studio of Duncan Grant. Doubly linked to the Stracheys as both the cousin and lover of Lytton, Grant had followed the family's political line by winning a prize for his suffrage poster 'HANDICAPPED!' (Figure 6.1). As described by the NUWSS journal, it showed two boats aiming for the Houses of Parliament, 'a stalwart Grace Darling type struggling in the trough of a heavy sea with only a pair of sculls, while a nonchalant young man in flannels glides gaily by, the wind inflating his sails'.[36]

Revelling in the freedom of her recently cropped hair, Ray spent a weekend in March 1911 at her aunt Alys's house, where the party included Bertrand Russell, the equally distinguished philosopher George Moore, Ray's sister Karin, and another Strachey brother—Oliver, who was thirteen years older than Ray. At that stage, his standing in the family was not particularly high: he had lasted less than a year at Oxford, had refused to follow the family tradition of working in the Indian Civil Service, and had failed to make a career in music; moreover, he had divorced his wife but insisted on retaining custody of their daughter. During the earnest discussions of Leibniz

Figure 6.1. Duncan Grant, 'Handicapped!', 1909.

and Hegel, Russell was preoccupied, wondering how he was going to tell Alys about his fresh love affair with Ottoline Morrell. Once the older generation was safely in bed, Oliver prepared a midnight feast of scrambled eggs with the Costelloe sisters and charmed them both.

Although Karin pretended not to care, it became clear within a few weeks that Oliver preferred Ray. Perhaps worried about his age, his daughter, and his financial situation, Oliver made no move. For her part, Ray was determined to ignore her mother's pleas to spruce herself up and hence secure her catch. She settled on a different future. 'I have decided I must go to London next winter for my engineering,' she wrote to her American aunt;

> The various people I have seen all advise it. I saw Mrs Ayrton, who is, as you know, a shining light in the world of electrical arc lamps, and she was most encouraging, & gave me a letter to the head of the Technical College, & he was also most kind. I am therefore spending this term in getting up the necessary things for the entrance examinations, and I begin work there in October.[37]

But Ray never did go to the Technical College. Within the next few weeks, her beloved grandmother had died, and she had proposed to Oliver and then married him. By the end of the year she was pregnant, living in India—and deeply disliking it. (Twenty years later, back in England, she campaigned for Indian women to be enfranchised, but her private reservations carried imperial overtones: 'I wonder what will happen if those very backward women get the votes...perhaps nothing worse than the men having it.'[38]) In February 1912, the couple sailed back to England, where they continued working optimistically if unrealistically on a history of seventeenth-century India that was intended to support them financially. Luckily for them, but not for Ray's stepfather, her mother secretly poured funds into both her daughters' households. Never one to waste time worrying about debt, housework, or social niceties, before long Ray was back in London and immersed once again in suffrage activity.

The Bloomsbury set is renowned for entangled sexual affairs, but although Ray certainly engaged in deep emotional relationships with women, it is hard to know whether these were physically realized. Many suffragists valued the close friendships generated by acting together in a common cause, and as Evelyn Sharp (sister of the folk-music collector Cecil) made clear, they forged unbreakable bonds that were 'only sometimes an intimate one'. 'You have a different feeling all your life,' she wrote, 'about the woman with whom you evaded the police sleuth and went forth...to be mobbed in Parliament Square or ejected from a Cabinet Minister's meeting.'[39] Ray kept great bundles of letters from Ellie—far outnumbering her own replies—who frankly declared her deepening love but worried that

> I cared more than you did and that has made me so repulsive and black hearted. It does make all the difference to me between Heaven & Hell to be with you....I am quite sure that unless *you* marry, I love you too much ever to marry; and yet there is very little chance of our setting up house for a long time to come....I miss you more & more.[40]

Bitterly jealous of Anna Shaw, Ray's older American friend, after the marriage to Oliver Rendel implored her 'for God's sake don't treat me as you treat the Rev. Anna....Don't let me down lightly; don't save my face....Be truthful at all costs. I am tougher than you think.'[41]

Ray and Oliver had two much-loved children, Barbara and Christopher, and she was content with the life she had forged. 'It is really ridiculous', she wrote, 'how I wish to fill my letters with shouts of joy and happiness. I am enjoying myself beyond measure, and like to say so again and again.'[42] Their London house was crammed with books (Henry James had long been her favourite author), threadbare Oriental rugs, and Oliver's piano; it also reeked of Ray's cigarettes. From around 1920 she spent much of her time in her equally untidy Sussex home, which she had built herself from compressed earth (hence the name it still carries—Mud Cottage) and was the only surviving evidence of her failed attempt to launch a society of women builders. Recruiting her friends to help, Ray laid bricks, installed plumbing, dug the garden, and reportedly enjoyed standing on her hands with no clothes on in the immense (and expensive) swimming pool. Her psychoanalyst sister Karin remarked (perhaps with a twinge of envy?) that Ray usually wore 'an amazing suit of corduroy breeches and coat which made her look like 100 elephants. Her disregard of her appearance is simply stupendous, she must be very sure of her inner worth, happy soul.'[43]

Within only a few years, Ray's marriage had settled into companionship. It was Philippa rather than Oliver Strachey who was her daily companion, and for holidays she chose to travel with women such as Mary Agnes Hamilton, a Labour MP and broadcaster. But although Oliver had three long-standing affairs, the couple remained emotionally close for over a quarter of a century, and she often turned to her 'old wiseacre' for advice and support.[44] Their only serious quarrels revolved around Oliver's first daughter Julia, who irritated Ray by still living at home in her mid-twenties, preoccupied with her appearance and her love affairs. When Julia went on a trip to Florence, Ray was relieved that 'the trail of scent, powder and fuss and telephone messages has left.... Poor Julia—I really think she must be half-witted in some directions.'[45] But she adored her own children, and they loved her deeply. 'I can't...wish to imagine a better mother than she was,' commented Barbara; 'She contrived to be that unique thing, a great woman, a truly wise parent and a very lovable person.'[46]

Although it seems that Philippa Strachey resisted Ray's suggestions of a closer relationship, they made perfect working partners: whereas Philippa excelled at foreseeing long-term consequences, Ray's impulsive extrovert character inspired immediate action. Enviably energetic, she always minimized difficulties and emphasized the positive: 'at any rate, *that's* to the good' became her catchphrase.[47] To judge from photographs, she only occasionally followed the advice she once put in a fictional heroine's mouth: 'What you *are* never gets a chance if you don't look right. You see, you've got to push against all the prejudices in the world in what you're doing; and you can't hope to get along if you pile up needless ones by the way you look.'[48] Woolf remained critical, describing Ray as 'floppy, fat, untidy, clumsy, and making fewer concessions than ever to brilliancy, charm, politeness, wit, art, manners, literature and so forth'.[49]

Contradicting Woolf's damning verdict, Ray played a crucial role in British politics and wartime activity. As well as engaging in the committee work her grandmother had predicted for her, she also went out campaigning: a close friend later recalled her 'standing up in a large car...looking just like Brunhilde, with her mop of then golden hair'.[50] Sometimes she bore the brunt of hostility intended for militant suffragettes: she told her family that at a talk in Greenwich she was 'pelted with mud and things, clothes completely ruined, dirty to the skin' and had to take refuge in a tea shop until the police told her it was safe to escape.[51] But the NUWSS policy of encouraging a combination of war work and peaceful protest was beginning to pay off. 'Hats off to the women of Britain,' wrote Arthur Conan Doyle after he visited a munitions factory in 1916; 'Even all the exertions of the militants shall not in future prevent me from being an advocate for their vote, for those who have helped to save the State should be allowed to guide it.'[52]

Internal Warfare

Now captaining activists instead of the Newnham cricket team, Ray found herself 'trying to harness a pack of wild elephants—and wild elephants with years of private conflict behind them'.[53] When war was declared, she endeavoured to manoeuvre her way diplomatically through angry discussions about patriotism and pacifism. Passions ran high: these women were fighting for their ideals. Although matters were complicated by controversy about forming an alliance with the Labour Party, there were essentially two main groups. Ray was among those who believed that it was every Briton's duty to drive the German army out of France and Belgium; their opponents felt equally strongly that as women, their mission was to oppose warfare, however justified the cause. With her American background, Ray was desperate to demonstrate her patriotic allegiance, confessing: 'I often think to myself with a real pang of pleasure that I have married an Englishman and have English children. And of course *now* that feeling has an intensity that it's hard to combine with tolerance—except that tolerance itself is an English thing.'[54]

Many of the Strachey relatives were conscientious objectors (COs), but—encouraged by Oliver, who had at last found his niche working for military intelligence in the War Office—Ray refused to be deterred by the abusive letters she received.[55] Privately referring to the pacifist adherents as 'the lunatic section', she felt that the NUWSS leader Millicent Fawcett did not fully understand the complexities of the situation and was voicing her own fervent patriotism as if it represented the views of all the members.[56] After an aggressive debate in April 1915, and much negotiation behind the scenes, many of the more vocal pacifists in the Society resigned. Ray wrote jubilantly to her mother that

> We have succeeded in throwing all the pacifists out.... It is a marvellous triumph that it was they who had to go and not us—and shows that there is some advantage in internal democracy, for we only did it by having the bulk of the stodgy members behind us.[57]

And that left Ray in a very strong position. Closely allied with Fawcett, but forty years younger, she exerted a powerful influence on NUWSS activities as their parliamentary secretary during a crucial period. By silencing extremists and refusing to readmit dissidents, she steered women towards two major goals: getting the vote and getting work.

Campaigning for the vote had paused for a couple of years while attention focused on the War, but in 1916, the government realized that there might not be enough voters at the next election because the men engaged overseas were ineligible. This gave suffragists an excellent opportunity to insist that women be included. After all, they argued, women had been running the country in the men's absence—and they were not afraid to hint at the threat of renewed militancy if their demands went unheeded. As Lloyd George put it in a parliamentary debate, 'the women's question has become very largely a war question'.[58]

Working closely with Philippa Strachey, Ray found herself absorbed by the negotiations involved in gaining the support of the press and politicians, and also in persuading her colleagues to accept Fawcett's uncomfortable but realistic compromise that only women over thirty should receive the vote. 'I find this sort of work almost too fascinating,' she enthused to her mother.[59]

> Pippa...and I have wound ourselves into the very thick of the intrigues, and we found ourselves dictating the leading article of The Times one day and sitting up until 4 a.m. preparing draft schemes for Mr Neville Chamberlain the next....I am packed so tight with interviews that I can hardly breathe, and find myself panting from one Cabinet Minister at dinner time only to breakfast with another the next day.[60]

Success came in sight on 19 Jun 1917, when the Bill for the Representation of the People Act was approved by the House of Commons. Ray described sitting amongst a crowd of women crammed together in the upper gallery behind a much-hated grille, peering through the 'bars of the absurd little cage in which they were penned [so] they saw chiefly the tops of the heads of the legislators'.[61] On the morning of the debate about the grille, 264 MPs discovered at breakfast a petition secretly signed by their wives: the glee with which Ray related this successful plot to get 'the oriental trappings' removed suggests that she was one of the conspirators.[62] Watching the final ceremonies in the House of Lords the following year, Ray jubilantly described how 'we all caught our breath a bit, and looked at each other...and then I skooted off in my car to tell Mrs Fawcett [who was] sitting up in a dressing gown by the fire—and we couldn't either of us believe it was true.'[63]

The electorate tripled overnight, enfranchising six million women. Ray's determination not to conflate pacifism with suffragism had paid off, because COs were disenfranchised for five years. On the other hand, after nearly four years of war, the national response was muted. Many agreed with the former suffragette Cicely Hamilton that when

> the measure enfranchising women became the law of the land a large proportion of those concerned would have been stirred to far greater emotion by the news that Big Bertha [a powerful German howitzer] had

exploded or that bacon had gone down threepence. And this lack of immediate enthusiasm or protest, so far from being reprehensible, showed a praiseworthy sense of proportion; food and the destruction of enemy armaments were of far more importance at the moment than the status of women and the ballot box.[64]

In any case, women were still not being treated equally. Because so many men had died in the War, women now easily outnumbered them—but by including an age limit, the new legislation ensured that men still had the majority voice. Even for a woman who was over thirty, there were further restrictions: she could only qualify if she were either a graduate voting in a university constituency or a member (or wife of one) of the Local Government Register. The vote might appear to be the government's reward for the women who had kept the factories going during the War, but most of them were too young for eligibility. As Sylvia Pankhurst protested, 'by its University and business franchise the Act still upholds the old class prejudices, the old checks and balances designed to prevent the will of the majority, who are workers, from being registered without handicap.'[65]

But only days after the Armistice, there was an unanticipated bonus: the 1918 Qualification of Women Act, which allowed women to stand for Parliament. By then, the electorate was less interested in the rights of women than in the rights of workers, and of the seventeen female candidates only one—the suffragette Christabel Pankhurst—campaigned on a woman's ticket. She lost narrowly to a supporter of the trade union movement, despite having rustled up support from the press and the government. Her skills at self-promotion are still evident a century later: one of the most celebrated suffrage photographs shows her wearing a light summer dress but supposedly casting her vote in the December 1918 election. In fact, the image shows her at a suffrage conference almost a decade earlier.[66]

In contrast, Mrs Oliver Strachey (as she was named on the posters) toed a delicate centrist line in her own bid to become an MP: 'I do not approve of extremes in politics,' she proclaimed in her manifesto;

'I distrust Revolution on the one hand and Reaction on the other.'[67] After years of internal wrangling in the women's movement, she recommended cooperation rather than confrontation: 'I believe with profound conviction that men and women should work together for the progress and good government of the nation, as they must for that of their homes.'[68] As an Independent without much money standing in a strong Conservative constituency, she had little chance of succeeding, although she did manage to win 43.5 per cent of the votes in the second of her three attempts.

The first woman to be elected to Parliament, the suffragette and Sinn Féin activist Constance Markievicz never went to Westminster. Britain's first female MP took her seat in 1919—Nancy Astor, a transplanted, unconventional American like Ray, and the latter who acted as Astor's unofficial adviser. A wealthy southern belle, and a strong Prohibitionist with a reputation for eccentricity, Astor was often condemned for being upper-class and out of touch, but with Ray's support, she remained in politics right through the interwar period. Four years later, there were eight women in Parliament, valued so highly that the nearest lavatory was quarter of a mile from their cramped offices.[69]

Working for Women

Suffragists are celebrated for winning the vote, but at the time many of them were more concerned about obtaining full professional status and equal salaries. In A Room of One's Own, Woolf described how she had inherited a legacy of five hundred pounds a year at around the time that women were enfranchised. 'Of the two,' she confessed, 'the money seemed infinitely the more important.'[70] For those not blessed with an aunt who had conveniently fallen off her horse in Bombay, economic independence had to be achieved through work. In the nineteenth century, inventions such as typewriters and telephones had opened up new opportunities for women to be self-sufficient, but—in a process that repeated itself time and again—had also reinforced the two-tier system of wages: 'As one process after another came to be regarded as "women's work",' Ray explained, 'it became, simultaneously, "unskilled", and the assumption that women must inevitably be routine workers, deserving of only a "woman's wage" remained unshaken.'[71]

As soon as war was declared, Ray dedicated herself to the central London Women's Service Bureau in Victoria Street. The birth of her second baby scarcely interrupted her. She had boasted for the first one that 'having babies seems to be very easy: I feel as if I were a living refutation of the "physical disabilities of women" argument!'[72] Insisting that women 'deserved equal pay, equal treatment and fair conditions', she immediately started finding ways to train women and place them in paid wartime work.[73] Already experienced in the administrative negotiations needed to coordinate volunteers, initiate projects, and wheedle money out of supporters—she reportedly 'concealed a hand of steel within the velvet glove'—Strachey spent much of the War chairing committees, enlisting royal patronage, and opening up unprecedented opportunities for women. As just one example, secure in her London base, she helped to run suffragist hospitals in Serbia. Ellie Rendel—by 1916 a final-year medical student—had volunteered to work on the Eastern Front, and she bombarded Ray with insider gossip. 'Dr H. [Alice Hutchison] behaved very badly out here,' she warned; 'she is a viper.'[74]

Although women were still regularly paid less than men for the same work, Ray initially felt that her strategy had succeeded: the War, she wrote, 'changed the wage-earning of women, almost overnight, from a disregarded or shameful business into a heroism'.[75] Yet she had the insight to recognize that this was, in a sense, a charmed and almost unreal period. Writing to her mother after the Armistice, she commented that 'we had all settled down to war as a permanent condition'—and she added presciently, 'perhaps we are a little afraid of peace'.[76] She was right to be concerned about the next few years, when returning soldiers reasserted their authority, squeezing women out of the positions they had held for the last four years, castigating them as upstart opportunists. As Ray put it, 'the very same people who had been heroines and saviours of their country a few months before were now parasites, blacklegs and limpets.[77]

But as she also pointed out, 'it is impossible to put the clock of experience backwards', and women's wartime experiences had a permanent effect on the entire nation.[78] For the next twenty years, Ray battled to retain the advantages that had been won, declaring that getting women into the Amalgamated Society for Engineers was 'quite the most important thing that could happen for women in England'.[79] Women fared worst in traditional fields, where a combination of female acquiescence, male assertiveness, and trades union activity ensured that men ousted women and reclaimed their previous positions. Women got on better in the newer trades—electrical products, light engineering—although their wages were often low.

Improvement was slow, but it did happen. With an eye on the new female voting power, MPs introduced a flurry of laws to improve women's rights, although in practical terms there was little immediate alteration in career prospects. Ray spelt it out: women were no longer disbarred from being lawyers, accountants, or civil servants—but gaining a qualification was very different from gaining employment. Teachers were allowed to continue working after they got married— but schools still had the right to dismiss them.[80] As the depression worsened, for many women paid work became a necessity, not a privilege, especially as they often had no option but to remain single in the unevenly balanced population.[81] Like Woolf, Strachey was financially independent, protected by her stepfather's allowance, which, she told him gratefully, 'makes all the difference to me between just scratching along, and being able to live as I like without perpetual cares'.[82] This privilege enabled her not only to campaign, but also to retreat into the solitude of Mud Cottage, where between 1923 and 1931, in addition to sitting on committees and orchestrating campaigns, she completed two novels, three feminist books, and an edition of her grandmother's papers.

In 1931, Berenson's investments collapsed and he cut off his allowance to both his stepdaughters. While continuing to write, Ray was forced to embark on a new experience—working for somebody else. Acquiring a full-time salaried post as Astor's political secretary, she refused to let herself be frustrated by taking orders, but instead tried to revel in the variety of new projects. 'I enjoy it more and more,' she wrote to her mother;

> I have been involved in (1) The politics of the BBC, (2) the forthcoming Children's Bill, (3) Allotments for the Unemployed, (4) Nursery schools in various parts of the country, (5) Pension cases in Plymouth, (6) Scandals in the Police Force, (7) the Unemployed Anomalies Act and (8) Civil Service Superannuation, today. And you can see that it keeps me hopping.[83]

Four years later, she achieved what must have seemed the perfect position: founding and running the Women's Employment Federation. And still she managed to deal with various family difficulties as well as escape on a long camping holiday in Greece.

Looking back two decades after the vote, Strachey judged that women could be particularly proud of having initiated major changes in two professions, the law and the civil service. Relatively rapid progress was made in the law, where women were (in principle, at least) treated equally and could serve as magistrates or on juries. The struggle for fairness in the civil service was more protracted, but with Parliament behind them, female pressure groups slowly forced that closed door ajar, if not fully open. In 1937, to their professed embarrassment, Ray and Philippa Strachey were guests of honour at a special dinner organized by the Council of Women Civil Servants—but Ray did admit privately, 'we *have* worked like blacks at it for 17 years'.[84]

Three years later, Ray Costelloe/Strachey died after an operation, aged fifty-three. As an outspoken and wilful reformer, she had aroused resentment in many of those she met, but nobody doubted her idealism. Forty years later, her travelling companion Mary Hamilton recalled the radiant expression on Ray's face when she proclaimed to the Carnegie Trust: 'I do not know whether I was able to convey to you the really passionate eagerness with which we long to do our work well.... [W]e are, in sober fact, enthusiasts.'[85]

Within a couple of years, Ellie Rendel and Virginia Woolf were also dead. The longest survivor was Philippa Strachey, who self-mockingly reported that she had become 'regarded as a sort of museum piece carefully assisted up and down stairs & provided with large arm chairs etc & have to play the part to the best of my ability & attempt to figure as Mary Kingsley, Emily Davies and Mrs Fawcett rolled into one'.[86] Ray Strachey was spared that decline into old age, remaining buoyant even in her last letter to her mother: 'we've all had enjoyable existences, packed with interest and ... good worthwhile Times, and been glad to have been alive.'[87]

PART III

CORRIDORS OF SCIENCE, CRUCIBLES OF POWER

7

SCIENTISTS IN PETTICOATS

Women and Science before the War

> But there is only one history. It began with the creation of man and will come to an end when the last human consciousness is extinguished. All other beginnings and endings are arbitrary conventions....The cumbrous shears of the historian cut out a few figures and brief passage of time from that enormous tapestry. Above and below the laceration, to the right and left of it, the severed threads protest against the injustice, against the imposture.
>
> Thornton Wilder, *The Eighth Day*, 1967

When she was one hundred years old, in 1972, the physiologist Mabel Fitzgerald was awarded an MA degree at Oxford. In a special ceremony attended by a couple of hundred people, the Vice Chancellor apologized: she should, he said, have received the accolade seventy-five years earlier.

Home-educated, Fitzgerald had defied convention by attending lectures at a time when many Oxford academics agreed that 'To be forced to teach ladies, who moreover are not members of the University, would deprive me of the chief interest I take in the efficiency of teaching.'[1] Although unable to graduate, Fitzgerald had gone on to publish several papers about her independent research. When the sympathetic scientist John Scott Haldane (uncle of the future Secretary of State for War) was planning a trip to investigate the effects of high altitude, he invited Fitzgerald to join the team going to America in 1911. Working on her own, she climbed through the mining mountains of Colorado to take blood and breath samples from residents in scattered encampments. After further similar expeditions, Fitzgerald began studying for a medical degree, but sacrificed her scientific career in 1915, first to carry out war work in an Edinburgh hospital and then to care for her sick sister. For decades, she lived quietly and unrecognized in Oxford.

Although Fitzgerald did eventually receive the tribute she deserved, countless other female scientists active in the decades before the First World War have been forgotten, perhaps because they lacked proper contracts or were employed on a temporary basis, or worked at home or allowed their work to be subsumed into a husband's. According to the official records, there were only five women at Imperial College when it was founded in 1907—two teaching staff and three students. However, the minutes of administrative meetings reveal repeated demands for more women's lavatories, which suggests that a substantially larger but concealed network of female workers existed.[2]

Some novelists used women's cumbersome clothes to symbolize the fetters binding their scientific heroines. A fictionalized version of the physicist Hertha Aytron declared she felt 'as though my work wore petticoats'.[3] 'Have you ever tried to run and jump in petticoats,' protested an imaginary biologist; 'Well, think what it must be to live in them—soul and mind and body!'[4] A real-life schoolgirl, Ida Mann, chose a different image. Taken to visit a training hospital, she gasped: 'Here was what happened when you studied for a degree in the University of London. Here was my life. Being a woman became only a minor physical disability, like a game leg; it need not be a deterrent after all.'[5] In the decades before the War, scientific women struggled to rip off such constraints both literally and metaphorically, but with only limited success.

Fitzgerald was as gratified by her influence on the future as by the official recognition of her talents. The day after her belated Oxford degree ceremony, she was invited to look back over her long life and decide what her most important achievement had been. From the pile of congratulatory messages at her side, she pulled out a letter from Sir Richard Doll, the epidemiologist who had linked smoking and cancer.

She had, he wrote, been the first to convince Oxford that women could do as well as men—and two out of three recent scholarships had gone to women. To adapt Bill Clinton's expression, women learnt to impress by the power of their example rather than the example of their power.

Hidden Heritage

As Jane Austen might have written, 'It is a truth universally acknowledged, that a single man in possession of a good scientific career, must be in want of a good scientific wife.' During the nineteenth century, most scientists carried out their experiments at home rather than in large central laboratories, often recruiting the women in the family to help—the most famous example is Charles Darwin. In closely knit scientific dynasties, aspiring young scientists married their colleagues' daughters, knowing that they had already been trained in experimental techniques. Well-educated and intellectually active, these wives became behind-the-scenes assistants, participating in their husbands' scientific research as well as undertaking crucial tasks such as editing and translating. Such female labour was only rarely recorded, and their hard work survives through achievements credited to their menfolk.

Eleanor Ormerod was determined to confront this male-dominated culture. In her position as government entomologist, every year between 1877 and 1901 she published a report on 'injurious insects' that attacked crops and plants—and she meticulously acknowledged the names of contributors who had sent in information from all over the country. Her own ambitions had been stifled not by a husband but by her autocratic father and her overbearing elder brothers. Encouraged by her sister Georgina, she embarked on her career only after the men in the family had died, converting her home into an all-female domestic research centre. Despite winning numerous medals from scientific societies, she was entirely self-taught, and felt her highest accolade was being awarded an honorary degree from the University of Edinburgh.[6]

The next generation of women could—in principle at least—go to university, and scientific research was becoming more important. Some of them, such as Dorothea Pertz, came from distinguished scientific families. Born in 1859, late enough to join the early intake of women studying sciences at Cambridge, and benefiting from her well-connected relatives, Pertz worked with the botanist Francis Darwin (son of Charles and Emma) but—like many of her female contemporaries—never had a formal position, even though she lectured at Newnham College. With Francis Darwin's backing, she published papers and became one of the first women elected to the Linnean Society. Operating in the margins, women like Pertz have received little recognition.[7]

Even women who played a substantial research role have been written out of the historical record. Concealed marital research partnerships continued for many decades. Men gained most of the credit—and marriage reduced the workload for a man while increasing it for his wife. Jeannie McGeogh had an MSc in chemistry, but after she married her research collaborator in 1929, he became a fellow of the Royal Society, while she went on helping him in the laboratory and *also* 'apparently without effort, maintained a high standard of social activity in the home'.[8] A London professor blamed the physicist Hertha Ayrton for her husband's death, complaining that if only she had 'put him into carpet-slippers when he came home, fed him well and led him not to worry...he would have lived a longer and a happier life and done far more effective work'. In actuality, Ayrton concealed the extent of her husband's illness by finishing his research and publishing it under his name.[9]

Couples integrated their careers and their marriages asymmetrically. After Grace Chisholm married her former tutor William Young, they worked together for three decades, producing around 200 mathematical papers as well as two books—and the first of these, on teaching children elementary geometry, was based on the education of their own son (who later died in 1917 when his aeroplane was shot down). Although they operated as a collaborative partnership, most of the papers appeared under his name alone. 'At present you can't undertake a public career', he explained to her; 'I can and do.' Their private conversations on the matter are unknowable, but Young's self-justification in a letter to Chisholm suggests that she had complained to him:

> The fact is that our papers ought to be published under our joint names, but if this were done neither of us get the benefit of it. No. Mine the laurels now and the knowledge. Yours the knowledge only. Everything under my name now, and later when the loaves and fishes are no more procurable in that way, everything or much under your name.[10]

Family responsibilities could eclipse the lives even of women with no husband or children to care for. When she was only nineteen, Dorothea Bate had talked her way into a position at the recently opened, all-male Natural History Museum. During the next fifteen years or so she trained herself to become an accomplished palaeontologist, and after three extraordinary journeys to the Balearics, shipped back crate-loads of fossils and typed up detailed scientific reports of her travels. But when her brother broke his leg playing polo, she was summoned to comfort their parents; effectively confined at home for around eighteen months, she saw her self-made career grind to a halt. Her parents then made what might seem an extraordinary decision. When her brother fell in love with a wealthy young woman, because of his lameness her parents refused to permit the marriage unless he handed over a substantial sum of money to her family. Although Bate's parents professed to despise the potential bride, they disinherited their daughter in order to finance their son. Dorothea Bate managed to survive this blow and eke out an existence through her scientific activities, but other women who received such setbacks chose to find a husband to support them instead.[11]

Occasionally somebody remembered to thank wives, although the tributes stressed the supportive aspects of their contributions. The bacteriologist Grace Toynbee worked collaboratively with her husband, but when he was appointed professor in 1888, the University of Dundee effectively demoted her by reporting: 'Any notice of Dr Frankland would be incomplete without some reference to Mrs Frankland, who has worthily aided and seconded him in his scientific career....' A Manchester professor presumably thought he was paying a generous tribute to his wife Gertrude Walsh, an organic chemist who had previously carried out her own research with Chaim Weizmann: 'Looking back, I can see how she subordinated her interests to mine, was always such a ready collaborator in scientific work, and cheerfully followed my chief vacation activity, namely mountaineering.'[12] Muriel Wheldale was already a widow when she was appointed a biochemistry lecturer at Cambridge in 1926, but even her obituary (by another woman) featured her paralysed husband: 'His great success in the face of incredible difficulties was in no small degree due to the encouragement and assistance he received from his wife. His triumph will always remain an inspiration to those who witnessed it.'[13]

In contrast, some men were so supportive that individual women were able to flourish. The radioactivity physicists Ernest Rutherford and William Soddy, for example, both had several female assistants and fully credited their contributions.[14] Even if cast as her husband's invisible collaborator, Ayrton owed her independent career to his protection and sponsorship. Moreover, unlike her friend Marie Curie, she could and did take out patents in her own right. Whereas Curie was legally a non-person until her husband was killed in a road accident, during the nineteenth century several Married Women's Property Acts meant that British female scientists owned their discoveries: in 1909 alone, 648 women applied for patents. While many concerned improvements in domestic devices, women were also filing patents for inventions related to aeroplanes, telegraphy, and marine engines.[15]

Many successful scientists picked their men wisely. The crystallographer Kathleen Lonsdale, one of the first women to be elected a fellow of the Royal Society, had intended to leave science after getting married, but her fiancé retorted that he had no need of a free housekeeper. She later advised women that if they wanted marriage and children, they should choose 'the right husband. He must recognise her problems and be willing to share them. If he is really domesticated, so much the better.' She also had an exceptionally encouraging research leader, William Bragg, who was so impressed by Lonsdale's stellar examination marks that he offered her a job and later found the funds for a childminder.[16]

Individual scientists such as Bragg could have a big effect. In 1890s Cambridge, the geology department had an unusually high proportion of female students because Thomas McKenny Hughes had taken advantage of a change in the rules to get married. His wife was thirty years younger than him, and he taught her geology so that she could become a researcher in her own right. Because she was deemed a suitable chaperone, women were allowed to join them on field trips, as well as attending mixed practical classes and lectures.[17] In contrast, in the physiology laboratories women were 'by no means generally acceptable—indeed, that is too mild a phrase...unacceptability not infrequently flamed into hostility. The woman student was rather expected to be eccentric in dress and manner; she was still unplaced, so far as the male in possession was concerned.'[18] Life was even worse for the handful of women at the Cavendish Laboratory, where physicists cultivated a romantic aura of male heroism pitted against unruly female nature. In this alien environment, researchers risked being decapitated by electric wires or choked by acid fumes. James Dewar is now celebrated for inventing thermos flasks, but he was so insistent on maintaining manly stoicism that two of his assistants lost an eye in the frequent explosions.[19] In contrast, a third of Cambridge's geological club were female. Louisa Jebb, for instance, was the first woman to win the University's agricultural diploma, and during the War she established the Women's Land Army, for which she was awarded the OBE.[20]

A particularly striking example of a male catalyst enabling female success was William Bateson, who gathered together a group of biologists at Cambridge between around 1900 and 1910. More than half of them were women, although for decades their significance for the future of evolutionary science was obscured, tucked away inside archives. Bateson is celebrated for coining the word 'genetics' and resuscitating the neglected experiments of Gregor Mendel on inheritance in peas. At the time, his ideas were regarded sceptically: odd as it might now seem, many experts thought that Mendel's theories were incompatible with Charles Darwin's evolution by natural selection. Struggling for recognition at the margins of scientific endeavour, and with a ready entrée to Newnham through his sisters and his mother, Bateson recruited a series of female graduates to his team. Regarded with contempt (or apprehension?) by many male academics, these women were not in a position to be too picky about a research topic. The advantage for Bateson lay in gaining some exceptionally smart and resilient researchers who could be employed more cheaply than men to carry out the same work.

Before long, Bateson's wife Beatrice had become enlisted at home, helping to record breeding data and to coordinate the project (she had insisted on marrying him, despite her parents' objections after he got drunk at their engagement party). He was an inspiring teacher who regularly invited students to

> a Sunday afternoon round of his own farmyard, garden, and poultry runs. As each cockerel and hen that he picked up had its inheritance demonstrated he would look round for some hand to receive the bird, till those of us with a country upbringing found our arms full of protesting poultry, and the procession would go round again returning them to their numbered pens.

But there were drawbacks: 'Some undergraduates in light Sunday suits were not very partial to this form of instruction.'[21]

Bateson's Newnham researchers studied a great variety of organisms— one specialized in guinea pigs, another in oats, a third in mice—but few of them achieved lasting fame. One who did was Edith Saunders (Figure 7.1); already established as an independent researcher in plant hybridization, she later became president of the Genetics Society. To her students, 'she seemed the embodiment of dedicated search for scientific truth. Rather austere in her tweed coat and skirt, with a very masculine collar and tie, yet with such a kindly twinkle in her eye.'[22]

Figure 7.1. Edith Saunders.

More and more women were attracted to the Bateson–Saunders duo, and as the experimental results began to look increasingly impressive, men asked to join as well.

National debates about heredity were mirrored by an internal Cambridge competition between the two women's colleges. While Newnham students were being drawn towards breeding experiments based on Mendel's approach, the women at Girton were being enticed to join a rival London programme based on statistics. An advertisement in the *Girton Review* condescendingly appealed for graduates 'of scientific tastes, leaving College for a life of comparative leisure at home [and] anxious to do some work which is useful and yet which need not necessarily involve hard or long-sustained brain work'. Several female recruits, including a genealogist and a meteorologist, joined the mathematical team at University College London.[23]

In her obituary, the biochemist Muriel Wheldale was commended not only for looking after her sick husband, but also for being one of Bateson's successful protégés. After entering Newnham in 1900 and gaining a first, she spent most of her life as a Cambridge academic and was a leading plant expert.[24] This was, of course, an unusual career trajectory for a woman, but she was not alone. Bateson's sister-in-law Florence Durham carried out a long-term study of guinea pigs and Mendelian genetics while she was employed as a pharmaceutical researcher. Like many other women with successful scientific careers, Wheldale and Durham benefited from belonging to a unique institution—the Balfour Laboratory. Opened in 1884 by Newnham in a converted Congregational chapel, for thirty years it enabled female life scientists to learn from and with each other.

The initiative to build the Newnham laboratory came not from a well-meaning benefactor, but from the students themselves, who launched their own fundraising campaign. Since the 1870s, Cambridge had been Britain's leading centre for the biological sciences, but female students resented the titters and stares when reproduction was discussed or a sample of brain tissue was passed round for inspection. Even though ten per cent of science students were women, their presence was barely tolerated: a Girton student later recalled how one of the star lecturers 'allowed us women to sit up in a gallery overlooking his big lecture room, full of men, among whom were not a few greyheads'. She also attended 'lectures on embryology one May term, in a tiny room, where men and women were squeezed together'.[25]

Squeezed together soon became squeezed out. As resistance towards women continued to harden, Newnham College decided there was no point waiting for the University to expand its facilities. First they created a laboratory of their own, and later they added a lecture room. Instead of wearing gowns like the men, the female lecturers adopted hats to complement their long skirts as a sort of uniform—'Not too large, turned up in the brim, just neat and appropriate'.[26] These women now had a physical space to study. Just as importantly, although students and academics discussed the very latest scientific ideas, they also developed a female subculture. Relieved from the disdainful male gaze, women at Newnham and Girton ran their own college science societies, gave their own lectures, and embarked on their own projects.

They also rebelled against the repressive attitudes of their predecessors. In 1897, the Principal of Newnham had expressed her 'good hope that women will do excellent work in the subordinate fields of science and learning, will do well much laborious work that needs to be done, though it is not very brilliant or striking, and will in particular prove excellent assistants'.[27] By the early twentieth century, such a limited ambition seemed old-fashioned. Led by Durham, the new generation of Newnham scientists successfully pushed for fellowships that would enable women to extend the bounds of knowledge by carrying out independent, original research.

United We Stand

Even girls from privileged backgrounds often received a skimpy education. Believing that they would never have to earn their own living, most pupils had little incentive to learn. Gender differences were perpetuated in a vicious circle that was hard to break. To protect their own male bastion, the civil service appointed female staff of a type unlikely to challenge traditional hierarchies, choosing them for their connections and warm-hearted enthusiasm rather than their expertise. The first woman chief education inspector was instructed that because most girls would become mothers, they 'need, in their upbringing in the schools, to be looked at from the maternal and physical aspect—not merely from what is called the intellectual and book learning aspect'.[28]

A high proportion of female science students came from the relatively small number of girls' schools that had made science a regular part of the curriculum by the early twentieth century. With a fresh emphasis on understanding rather than rote learning, educational reformers began to insist that children should be trained practically in a laboratory, not just by studying books. Richer girls' schools invested in new buildings and hired university graduates, who wore a cap and gown to teach their classes. Presumably it was one of them who, in an inspired move for funding to improve facilities, advertised the risks of being without a properly protected laboratory by leaving an open flask of chlorine in a classroom just before a tour of the school's administrative committee.[29]

Unconventional and determined, female students made a large impact relative to their numbers. In 1900, only 3,349 women were at Britain's twenty-two universities, over 500 of those at the two largest, Oxford and Cambridge. Mainly from more privileged backgrounds, these Oxbridge women have left greater traces because the individual colleges kept detailed records. Between 1881 and 1916, 403 women studied science at Cambridge; impressively, almost 10 per cent took up university careers, even though they were not allowed to graduate until half a century later than at London and the other new universities.[30] For a short period after 1904, when Trinity College Dublin first went co-educational, around 700 Oxbridge women managed to find a way round the ban by taking a fast mail boat to Ireland. In an unusually liberated move, it had been deemed unfair that although the College would soon be awarding degrees to women, Irish students currently at Oxbridge were unable to take advantage of that option, and they were allowed to graduate in Dublin. Some of these graduates went on to have distinguished scientific careers, such as Margaret McKillop, the first woman to get a first in chemistry at Oxford, and Annie Homer, a Newnham biochemist who specialized in antitoxins but changed track in the 1920s for a career promoting the development of mineral resources in Palestine.[31]

Female science students were often mocked, excluded from classes, and provided with inferior accommodation. Even sympathetic academics could sound condescending. 'I think you would be amused,' the world-famous Cambridge physicist J. J. Thomson wrote to a female friend in 1886, 'if you were here to see my lectures—in my elementary one I have a front row consisting entirely of young women and they take notes in the most praiseworthy and painstaking fashion, but the most extraordinary thing is that I have got one in my advanced lecture.'[32] When she was at Girton studying mathematics, Grace Chisholm carefully avoided being taught by Mr Young, who had a reputation for making girls cry. Her friends had warned her about 'Poor Ethel!...She went all wrong with dimensions and Mr Young talked about her in the third person and held her up to ridicule.' When she was invited to attend the lecture of a famous professor, Chisholm had to persuade the head of the college not only to give her permission for the couple of miles' walk into Cambridge but also to write a special note for the professor.[33]

Among the academic subjects, science posed particular problems, partly because of the practical classes (especially in biology), but also because of the ingrained belief that women think intuitively rather than rationally. Hidebound educators agreed with the view expressed in 1914 that 'it has long been obvious that girls require a course in science quite different from that which it has been customary to provide for boys.'[34] No wonder, then, that one of the most successful fundraising initiatives was for a new course in 'home science' at King's College for Women, a thinly disguised attempt to heighten the status of domestic drudgery. Over £100,000 (estimated at more than £10 million in its modern equivalent) was raised, but as one critic fumed, all over the country there were intelligent girls 'crushing their rage in the folds of dish-cloths' who could have entered 'a new world' if the money had been spent on scholarships.[35]

Funding presented a major obstacle for would-be students lacking rich and indulgent parents. Margaret Benson taught for seven years before, at the age of twenty-eight, she had saved up enough money to enrol at University College London. After getting a first-class degree in botany, she won a research scholarship and then spent most of her life heading the botany department at Royal Holloway College. On top of creating their botanical garden, museum, and herbarium, she specialized in the new technique of studying fossil sections with a microscope. Working in a garden shed, she prepared her own slides with a gas-powered cutting machine, and made important contributions to the study of fossil plants.[36]

Arriving at university with a good grounding in science, some women started to outperform the men. After successfully cramming two years of study into one at University College London, Marie Stopes sent a jubilant letter to her mother Charlotte, a suffragette and Shakespeare scholar. Underlining key words in her frenzy, Stopes crowed:

> The most unlooked for news is that this year I have got my degree. I am now, BSc. Not only have I got it I have got it very well, I have got 1ˢᵗ Class Honours in Botany…also Honours in Geology 3ʳᵈ Class and am the only candidate with honours, the others (men only) all failed, so my name stands alone in the list. As it is supposed to be impossible to take one honours in a year, to get 2 is very nice.…[37]

For women like Stopes who managed to stay the course and win a degree, pursuing a scientific career demanded leaping—or limping— over yet more hurdles. Analysing graduate destinations can make depressing reading: before a sudden blip in 1914, the tables show that teaching was the by far the most viable career option for women.[38] As one jaded scientist put it, 'they learnt science to teach women what to teach other women who in their turn taught others.'[39] Worried about their future careers, many female students dissociated themselves from political activism. Although there were suffrage societies in most universities before the War, meetings were often disrupted by male vandalism, and much depended on the energy of individual enthusiasts such as Ray Costelloe and Ellie Rendel.[40]

Resilience rivalled brilliance as an essential qualification for earning a scientific living. A female lecturer advised:

> The woman pharmacist has something else to weigh besides expense: it is the question of her sex, and the fact that at present she is a pioneer in her profession, and must naturally turn to where she thinks an enterprising woman will be respected and her ability made use of.… [T]he places to avoid are centres, such as cathedral towns, where anything new is looked upon with suspicion, and must stand the test of time before it can be trusted.[41]

Landing a university position was also hard. During the four decades before 1910, there were only fourteen female researchers at Cambridge's Cavendish Laboratory. In London, on learning that the physiologist Harriette Chick had applied for a place at the Lister Institute of Preventive

Medicine, two researchers immediately implored the Director not to commit the folly of appointing a woman to the staff. Luckily for her and for science, their pleas were ignored: she was appointed on equal terms, and remained at the Institute for the rest of her life. During the War, she supervised a group of women researching into diagnostic techniques for typhoid, dysentery, and other infectious diseases that were decimating troops overseas. Without Chick and her female colleagues, the Institute would have ceased to function while the men were away.[42]

Even so, evidence of outright misogyny can be ambiguous and thus misinterpreted. Historians routinely cite a conversation in which Lord Haldane was informed by the all-woman Bedford College: 'They do not go to Professor Ramsay.... He does not encourage women to work with him particularly. I think I am not misstating the fact that he rather discourages women in his laboratory for research purposes.'[43] Perhaps sometimes he did, but he was also an early campaigner for women to be allowed full membership of scientific societies, and was even cited in an Australian article of 1911 for praising women.[44] Ramsay did, in fact, have two female research assistants, although he left no written record of them himself. A visiting German physicist, Otto Hahn, reported that

> One of them, a Miss O'Donaghue, was especially well disposed to me. I had once shown her preparations in the darkroom, where they were fluorescent even in the dim light, and since then she always liked coming into the darkroom with me. But I never dared to kiss her. I had been warned that in England a kiss was tantamount to an engagement, so that it might lead to one's being sued for breach of promise.[45]

It seems that Ramsay recognized individual excellence, even if its owner was wearing a skirt. Possibly he was worried not that women were intellectually inferior, but that they were too distracting and might leave as soon as they got married.

A doctorate provided no guarantee of employment, let alone a rung anywhere near the top of the academic ladder. England's first woman professor, Edith Morley (English Literature at Reading: the second was the physiologist Winifred Cullis),[46] observed from her own bitter experiences that although in principle university posts were open to both men and women, in practice preference was often given to men. Like many other women, she was particularly aggrieved when women-only colleges appointed men to senior positions. She imparted sound advice that regrettably still needs reiterating: 'This fact should not deter *fully qualified* women from applying.... The power of suggestion is very great, and it is well to accustom appointment committees to the consideration of women's claims.'[47]

Those women who did secure a university position were often engaged in back-room research. With a narrower range of options than men, women were frequently channelled into time-consuming tasks of administration and teaching—and inbuilt pay differentials meant that experienced women could be hired more cheaply than men, and so often ended up doing lower-grade work. Even supportive professors generally judged women unfit to address audiences of male undergraduates or to run their own projects; instead, they restricted their paternalism to ensuring that assistants were reasonably well rewarded for off-stage work. At Imperial College London, during the First World War, Muriel Baker—wife of the professor of inorganic chemistry—was isolated at home in a small laboratory where she experimented on using stockings stuffed with cotton wool to absorb poisonous gases.[48]

Whatever the official policy of an organization, in practice women were mostly excluded from those informal discussions at which key decisions are made. After a prolonged campaign at University College London, some female academics eventually managed to secure their own tea room—a former chemistry laboratory, with a gas burner installed in the 'stink-cupboard'—but Margaret Murray reported sarcastically that when she was invited into the men's common room there was 'such a sanacker-towzer of a row that even [her male ally] quailed, and a feminist invasion was averted!'[49] Marie Stopes experienced similar obstacles. 'Women high up in scientific positions, women with international reputations,' she protested in 1914, 'are shut out from the concourse of their intellectual fellows.'[50]

Like many women—Marie Curie, for example—Stopes had chosen to enter a new field, palaeobotany, and she became the first female lecturer appointed at Manchester University. Also like Curie, she benefited from progressive parents (her suffragette mother was exceptionally pushy) who ensured that she had an excellent scientific education. Regarding herself as a child of the British Association for the Advancement of Science, where both her parents gave lectures, Stopes travelled on archaeological field trips with her father and gave her first talk on fossils as a schoolgirl. Yet despite this evident ability, she felt marginalized.[51] Although she had received a travel grant from the Royal Society to carry out research in Japan, when she was invited to display some of her specimens she declined after being told that she could not present them at the Society herself, but must send her butler or a male student instead.[52]

Women were formally banned from many professional scientific societies, although attitudes varied. Results often depended on local initiatives by a few individuals. When the Royal Society refused to admit Ayrton, the fellows hid behind the terms of their existing charter, although it would have been straightforward to add an additional clause if they had wanted to. That was exactly the step taken in 1915 by reformers at the Royal Astronomical Society and the Physiological Society. By then, the Royal Institute of Chemistry had already allowed women to join, although accidentally rather than intentionally: in 1892, a wily London graduate called Emily Lloyd sat the Institute's qualifying examination under the name E. Lloyd—and when she passed it successfully, the Institute could find no way of wriggling out.[53]

With no old school tie network to fall back on, women formed their own connections, found their own ways to raise funds, and created their own centres for female academics. At the all-woman Bedford College, for example, the geologist Catherine Raisin insisted on reaching for the highest academic ideals, declaring in 1901: 'This will be for women (as in the older universities it has been for men) to encourage the desire to search for knowledge for its own sake.' A hardened traveller and often the only female delegate at conferences, she was an inspiring teacher, and during her thirty years at the College, the number of science students went up fourfold.[54]

Male resistance remained high. When the Biochemical Club was founded in 1911, the members initially voted to exclude women. Only two years later, they had yielded to pressure from campaigners, who included two former Newnham women, Ida Smedley and Muriel Wheldale. After an extraordinary series of petitions, votes, and underhand manoeuvres, the rules remained unchanged until 1919, when the Sex Disqualification Act made it illegal to exclude women (whether married or not) from a society or a profession.[55] But of course there was still no *obligation* to include women. Although in 1922 the Women's Engineering Society reminded the Royal Society of the change in the law, the first two female fellows were not elected until 1945.

Inspired by hearing Millicent Fawcett speak in Manchester, Ida Smedley launched the British Federation of University Women (BFUW). Its ringing manifesto set out a fourfold mission: to enable collective action, to encourage independent research, to facilitate cooperation, and to find places for women in public life.[56] One of its first presidents was Ethel Sargant, an early natural sciences graduate from Newnham who had spent a year learning about botanical research at Kew Gardens, but then—like so many women (in the arts as well as the sciences)—found herself operating from the fringes of academia. With no formal position, she continued to work at her own laboratory in her mother's garden and later at home in Cambridge. Although she was made an Honorary Fellow of Girton, taught students in her own laboratory, and published many papers, she never enjoyed a professional career.

Unmarried, Sargant presumably lived off inherited money, but during the War she spent much of her time compiling for the Board of Trade a list of university women willing to do war work. In a handwritten note, she recorded her successes, which included placing three mathematicians in a naval aeroplane construction station as draughtsmen [*sic*], three munitions supervisors, a laboratory steward in the Admiralty, an assistant mechanical engineer, an assistant at the National Physical Laboratory, and a botany lectureship in South Africa.[57]

Growing from an initial kernel of seventeen in 1907, by 1930 three thousand academics had joined the BFUW, all frustrated at the difficulties of getting research funding, promotion, or even recognition. By then of course, the War had intervened, when many female scientists had put their own interests on hold in order to work for their country. Afterwards, the high number of fatalities meant that potential husbands were in short supply: a science degree no longer represented a stepping-stone to marriage, but was valued instead as an entry ticket to a money-earning career.

8

A SCIENTIFIC STATE

Technological Warfare in the Early Twentieth Century

But it was partly the simple, pathetic illusion of the day that great things could only be done by new inventions. You extinguished the Horse, invented something simple and became God! That is the real pathetic fallacy. You fill a flower-pot with gunpowder and chuck it in the other fellow's face, and heigh presto! the war is won. All the soldiers fall down dead! And You: you who forced the idea on the reluctant military, are the Man that Won the War.

Ford Madox Ford, *No More Parades*, 1925

'The lamps are going out all over Europe; we shall not see them lit again in our life-time.' The Foreign Secretary Edward Grey apparently made that remark to a colleague on the eve of the First World War, gazing down from his window at the gaslights being illuminated in the street below, knowing that he was himself going blind.[1] A hundred years later it is the only thing that most people remember of him. It became his most famous utterance not because it was necessarily true, but because it evoked such a nostalgic memory of an old and civilized order that had been rudely brought to an end by the disruptive forces of warfare.

Such sonorous pronouncements get repeated so many times that they start to acquire the ring of unchallengeable truth. But there were and still are many different ways of thinking about the state of the nation in the early twentieth century. Not everybody shared Grey's view that a tranquil era had been unexpectedly disrupted. Many people regarded Edwardian England as already a turbulent epoch. Compared with much of Europe, Britain was technologically advanced, and romantic techno-phobes denounced increasing industrialization for destroying not only the nation's idyllic countryside but also its moral code—financial gain, they argued, was becoming more highly valued than freedom and personal integrity. War was still four years away when one of E. M. Forster's heroines sadly contemplated the artificial red glow of petrol-laden London encroaching like rust across the rural landscape; life, she lamented, would be 'melted down, all over the world'.[2]

Other sectors of the population were more bellicose. Pointing to the statistics of rising unemployment, poverty, and sickness, some Darwinians argued that war was needed to cull the human population and so ensure that only the strongest survived. The poet Rupert Brooke, then in his twenties, welcomed war as a purifying agent that would cleanse the nation and energize its youth:

> With hand made sure, clear eye, and sharpened power,
> To turn, as swimmers into cleanness leaping,
> Glad from a world grown old and cold and weary...[3]

In 1913–14, while suffragettes were slashing old masters, the Italian Futurist Filippo Marinetti was lecturing in London about the glorious-ness of war. Artistically, his British peers were Percy Wyndham Lewis's Vorticists—often called the shock troops of British Modernism—who were intent on revolutionizing art to celebrate technological achieve-ment. '[Y]ou wops insist too much on the Machine,' Wyndham Lewis told Marinetti; 'We've had machines here in England for a donkey's [*sic*] years. They're no novelty to us.' The month before Britain declared war against Germany, the Vorticists opened hostilities against 'Clan Strachey' and establishment targets by publishing *Blast*. With its puce-pink covers and stark black sans-serif typography, the journal was a call to arms for Britain's artistic community.[4]

Retrospective reactions have been equally varied. Over the last century, the messy actuality of events between 1914 and 1918 has largely coalesced into myth-like narratives of heroism, self-sacrifice, and administrative blundering. Especially in films, dramatic death scenes take precedence over the unremitting misery of daily muddy squalor. Sanitized and dramatized, scripts exploit the same grisly fascination that lures clusters of voyeuristic spectators to gawp at gory motorway accidents and aeroplane crashes.

The wartime realities experienced by many participants remain scantily commemorated because they do not fit standard formats. Not everybody—not even the most highly trained soldiers—spent all their time fighting: long stretches of boredom had to be endured when there was very little to do, sometimes with no obvious reason provided for the lack of action. British discussions of the War are overwhelmingly dominated by events on the Western Front, and relatively few consider what was happening across Europe and out beyond its borders into Africa and the Middle East. And although the predominantly middle-class volunteers who signed up for the Voluntary Aid Detachments (VADs) undoubtedly played an important role, their demure white uniforms fit far better with wishful-thinking images of women's contributions than do the greasy overalls and mindless, menial tasks of the far greater numbers from less privileged backgrounds working in factories at home.

In many persistent versions of the early twentieth century, Britain features as a backward nation headed by an unprepared government. According to that set of stories, the country was suddenly and reluctantly forced into a flurry of scientific innovation to cope with a war that nobody had anticipated. One way of refuting that is to examine how language altered. In a small Essex village, a church rector spent the War meticulously gluing newspaper clippings into his compendium of 'English Words in War-Time'. He collected around 4,000 terms, but most of the technological ones—aerial raider, bombdropper, Zeppelinophobia—described wider uses of older equipment rather than the introduction of brand-new inventions.[5]

Britain was already a scientifically and technologically advanced nation, and had already been taking pre-emptive measures against the outbreak of hostilities. In particular, the Navy had been leading competition against Germany by procuring the world's latest and largest vessels. To maintain the country's imperial power structure around the globe, experimental development programmes had long been underway in industrial laboratories, military arsenals, hospital departments, agricultural stations, arms factories, and other non-academic organizations, often funded privately by commercial enterprises or philanthropists. In addition, many university projects were geared towards practical goals and financial gain just as much as the accumulation of knowledge.[6] Although many members of the public may have been taken by surprise—the maternity hospitals were packed with premature births in the first half of August—government, military, and industrial organizations were well prepared for a short, sharp war. Contradicting the mythology, Britain was not a Victorian nation unexpectedly forced to engage in a twentieth-century battle.

Scientific Battles

> Some one had blunder'd:
> Their's not to make reply,
> Their's not to reason why,
> Their's but to do and die...

The Charge of the Light Brigade gratifyingly confirms ingrained convictions that military leaders are not only incompetent but also oblivious to the fate of ordinary soldiers. Cynics looking back after the First World War added scientific ignorance to the mix: 'Fortunately for mankind, the army has usually been the refuge of third-class minds...Hence the paradox in modern technics: war stimulates invention, but the army resists it!'[7]

Whatever evolutionists might say about natural struggles for survival, in human warfare the casualties are predominantly healthy young men. In the First World War, a very high proportion—a fifth—of Oxbridge men were killed. Surely, runs the complaint, the cream of British youth should not have ended up as cannon fodder? Should scientists not have stayed in their laboratories to invent more effective weapons? William Lawrence Bragg received a Nobel Prize in

1915, so why was he allowed to join the Horse Artillery? Fortunately, he survived to head Cambridge's Cavendish Laboratory after the War, where he made a special point of encouraging female scientists, recognizing that to have got thus far, they must have been unusually clever and determined—and also willing to work for less pay and poorer chances of promotion than men.[8]

In contrast, the brilliant X-ray physicist Henry Moseley was sent to Gallipoli as a telecommunications officer, where he was hit by a sniper. His parents received an official letter that was presumably meant to console them: 'your son died the death of a hero....He was shot clean through the head, and death must have been instantaneous.' In her diary, his mother wrote simply 'My Harry was killed in the Dardanelles.' The country's leading atomic physicist, Ernest Rutherford, had been Moseley's research supervisor, and he campaigned for scientists to be deployed more usefully, pointing out that 'his services would have been far more useful to his country in one of the numerous fields of scientific inquiry rendered necessary by the war than by exposure to the chances of a Turkish bullet.'[9]

That poignant version is factually correct, but it leaves out Moseley's complicity in his death. Moseley was—like Wilfred Owen—an early volunteer, a patriotic young man so determined to enlist that he turned down the offer of a new scientific position. Aware of his poor chances, he wrote a will while overseas leaving everything he owned to the Royal Society. In 1915, months before compulsory conscription was introduced, a deputation of eminent chemists visited the Board of Trade, and from then on, many scientific researchers working on war-related projects were exempted from service.[10] The following year, a list of reserved occupations was established to ensure that those with essential skills remained at home rather than being forced to fight.

When war was declared, most people assumed that it would soon be over. In particular, scientists insisted that they belonged to an international community transcending national differences. Research would, they declared, continue as usual—and thanks to all those compulsory classics lessons at school, they could converse with each other in Latin if necessary.[11] But foreign friends were soon being converted into bitter enemies. Scientific journals began carrying regular reports of promising researchers who had been killed, and the mounting tally forced older survivors to change their views. In what must have seemed like another life only a few months earlier, they had shared jokes, meals, and experimental results with German colleagues they had known for years; now, they openly expressed their enmity. Despite the protests of pacifists, the British scientific community agreed that Germany should be excluded from international projects.[12]

The First World War is often dubbed 'the chemists' war', but the only truly new military invention of the War had nothing to do with chemistry—it was the tank. Poisonous gases had already been introduced, although using them violated the Hague Convention, the gentlemanly agreement initiated in 1899 to ensure fair play in the sport of war. After the Germans first used chlorine in April 1915, both sides went on to devise ever more noxious varieties. The aeroplanes, battleships, machine guns, and submarines that had previously been developed by engineers and physicists were just as significant as toxic gas and explosives. When he wanted to satirize military strategy under the ironic caption 'Wonders of Science', the political cartoonist Will Dyson made no comment on gas warfare; instead, he drew aircraft piloted by apes bombarding a city from the air, a new and crucial arena right from the outset.[13]

Old technologies generally stay in use long after new ones have hit the headlines. One singularly effective device owed nothing to science and was very simple: barbed wire, originally invented in America to prevent cattle from wandering outside their grazing grounds. Like many soldiers, when the poet Siegfried Sassoon came home on leave, he bought the supplies he would need to take back with him, including two pairs of his most essential equipment—barbed-wire cutters. If military success is measured by the number of enemies killed, including civilians as well as soldiers, then by far the most important weapon in the War was neither the machine gun nor poison gas, but the rifle. Battles were won by combining improved versions of old weapons with information gathered through intelligence, especially reconnaissance from the air. Similarly, horses were the most common form of transport: unlike motorized vehicles, they could get across muddy fields and, as long as they were kept well-fed, never broke down. At the beginning of the War, the Army requisitioned 170,000 horses from farms and hunting stables, while thousands more were shipped in from all over the world for delivery to war zones.[14]

Edwardian Britain is often affectionately portrayed as a nation locked in the Victorian age; its industries in decline, its universities committed to studying the past, and its military forces sadly bereft of modern technology.[15] If that were truly the case, then why did one of the twentieth century's greatest intellects—the Austrian Ludwig Wittgenstein—choose to leave Berlin in 1908 and study for his doctorate in aeronautical engineering at the University of Manchester? The answer is simple. The country was not industrially backward: in particular, it did not lag behind Germany, and there was close interaction between the government, the military, industry, and academia. An important centre for aeronautics, Manchester boasted some world-famous professors and could attract high-calibre students such as Wittgenstein, who won a research award for designing a propeller with jet engines on the tips of its blades.

Even today, few Members of Parliament are trained scientists, and a century ago, many of the highly educated men in power knew more about Electra than electrons. Yet despite not themselves being actively involved in research, some major leaders were convinced that technological solutions were essential for maintaining Britain's industrial, military, and imperial lead. In 1905, Richard Burton Haldane, a professional lawyer with a keen interest in science, became Secretary of State for War, and he set about reforming the British armed forces to make them more scientific. During the War, the government poured money into improving existing technology and medical techniques.

For example, David Lloyd George recruited scientists to advise the Ministry of Munitions, and when attempts to develop a tank ground to a halt, Winston Churchill intervened to spend Admiralty funds on hiring engineers and making prototypes.

Viewed from the perspective of scientists, the War was beneficial not so much because it financed a wave of innovation, but because it helped them negotiate more powerful positions for promoting research, education, and investment. In 1902, there were fewer than 2,500 scientists in Britain; after the War, the total was increasing by that number every year. This boom stemmed partly from scientists' ability to organize themselves into effective pressure groups for convincing the government and the military that science was crucial. The Royal Society was trying to manoeuvre itself into a position of power and take control of deciding how public money should be allocated for funding national science. To boost their collective status, eminent scientists emphasized how important their contributions would be to Britain's wartime strategy. For decades, scientific campaigners had been pushing for greater recognition, arguing that rationally trained men made better citizens and that industrial research would benefit the nation. With the outbreak of war, they seized the opportunity to step up their campaign by infusing earlier debates with a sense of urgency. 'In this hour of national emergency,' declared an editorial in *Nature*, 'there is no time to be lost. We cannot all be soldiers, but we can all help, we men of science, in securing victory for the allied armies.... [T]he sooner we co-operate for the good of the nation, the sooner will the war be over.'[16]

The author of this patriotic but self-serving appeal was Sir William Ramsay, a Nobel Prize winner in chemistry, whose reputation had been somewhat tarnished by a misjudged project to extract gold from seawater. His jingoistic language reveals how much attitudes have changed during the last hundred years. Reminding his readers that it was their responsibility to make sure the rest of the world was moulded into 'the Anglo-Saxon character', he claimed that recent scientific success owed little to Germany, so that 'the restriction of the Teutons will relieve the world from a deluge of mediocrity.' Warming to his theme, he continued to denigrate the German nation: 'Much of their previous reputation has been due to the Hebrews resident among them; and we may safely trust that race to persist in vitality and intellectual activity.'[17]

By November 1914, senior fellows of the Royal Society had formed a War Committee to recruit university colleagues from across the country and advise the government on investigating military applications of their research. For their part, the armed services were already major sponsors of technological innovation and so naturally professed themselves eager to recruit additional scientific expertise. A few months later, in April 1915, the Royal Navy expanded its research programme still further by setting up its Board of Invention and Research (BIR), which invited Royal Society scientists to help them decide how funds should be allocated to new projects. The Board's scientists were also asked to assess proposals that were flooding in from the public—41,000 of them in three years. Unsurprisingly, many of the inventors were inspired more by the promise of a reward than by any creative genius, and very few feasible suggestions were received. One of the most memorable contributors was Harry Grindell Matthews, a controversial, flamboyant engineer who claimed in the 1920s that he had invented an electric death ray. After demonstrating his remotely controlled boat on a pond in Richmond Park, he convinced admiralty officials to award him a substantial prize, but this anti-Zeppelin equipment was never actually used.

Especially after conscription was introduced, many academic scientists were allocated technical tasks that took advantage of their civilian expertise—geologists were deployed to dig tunnels, geographers to draw maps, physicists to develop communications systems, chemists to superintend explosives. Unlike military and industrial laboratories, university departments started to empty, or rather to change: the temporary absence of men gave female scientists an unprecedented opportunity to take up more prominent positions. For instance, at Imperial College, a third of the students and sixty members of staff had signed up by the end of 1914, and women seized the opportunity to give lectures, attend classes, and carry out more challenging research than before.

Research scientists who stayed behind—women, the elderly, men classed as unfit—put peacetime projects to one side and instead concentrated on improving military equipment and medicine. The new redbrick universities took advantage of the War for demonstrating their national importance, diverting their programmes to match national wartime requirements. As just a few examples, Birmingham chemists collaborated with Lever Brothers to develop a powerful tear gas that could penetrate the filters in German masks, Manchester physicists carried out sound experiments to improve submarine detectors, Bristol pathologists investigated industrial fatigue in the munitions factories, and London chemists devised a more effective filling for bombs and grenades (codenamed SK for South Kensington).

Government Machines of War

Before the War, the scientific world was, of course, overwhelmingly male, but it was thanks to a woman—Lady Jane Taylor, the widow of a distinguished general—that in 1908 the British government received an urgent telegram: 'Aeroplane primarily intended war machine stop.'[18] This missive arrived because after three years of commercial negotiations with Britain's War Office, the Wright brothers had just completed their first successful flight in France. Wilbur and Orville Wright may have had their heads in the clouds, but when it came to business, they kept their feet firmly planted on the ground. To promote their financial interests internationally, they appointed a wealthy New York businessman called Charles Flint, who immediately engaged a British agent—his well-connected friend, Lady Jane. Promised a commission on any aeroplane sales, she immediately started putting pressure on some of Britain's most influential aristocrats, including the newspaper proprietor Lord Northcliffe, and Haldane, Secretary of State for War.[19]

After some tetchy exchanges, Haldane rejected her proposals—not because he was too blinkered to understand the benefits of flight, but because he needed to maintain the official secrecy surrounding Britain's own aviation research projects. Although not a scientist himself, his younger brother John was an Oxford physiologist who later served on the Royal Society's War Committee, and Haldane believed that science paved the path to the future. One of his first steps on assuming power was to rationalize the army's organization, and next he turned his attention to aeroplanes. In his view, this recent invention was too important to be left in the hands of engineering entrepreneurs, and Haldane created a government advisory committee stacked with eminent academics. Instead of limping along through trial and error, British aviation would—he declared—flourish under a rigorous scientific strategy.

Not everyone agreed with Haldane's commitment to scientific rationality. Lord Northcliffe was just one among many critics who insisted that practical experience outweighed scientific theorizing, and by 1914, his *Daily Mail* had handed out generous prizes to independent pioneers (by some calculations the equivalent in today's money of over £2 million). The government and the military were setting up more and more committees manned by academic scientists, but this tussle between advocates of hands-on knowledge and of scientific expertise continued throughout the War.

Britain was ahead of the game in the early twentieth century, but science enjoyed nothing like the professional prestige it does today. The word 'scientist' was only edging into standard English, and journalists mostly used terms such as 'men of science' or 'scientific men'. Even the meaning of science seemed fuzzy. One eminent medical professor confronted the problem, offering doctors a working definition that pragmatically emphasized its value for society: science, he said, 'is a Latin word for the best way of doing what is to be done'.[20] Broadly speaking, there was a divide between researchers elucidating the laws of nature—'pure' scientists as they liked to call themselves— and those of a more practical bent who felt they were improving the world by solving problems. While many academics preferred to regard themselves as being above the sordid world of commerce, more entrepreneurial inventors filed patents to protect their financial interests. For purists, establishing intellectual property rights in this way contravened the scientific ideology of freely exchanging information.[21]

Aviation was split between experienced engineers, many of them flying their own aeroplanes, and scientists who may have been experts in aerodynamics, but were criticized for knowing little about the day-to-day difficulties involved in keeping a machine up in the air. In 1914, rivalry between the two camps persisted, and mutual antagonism flared. An MP who chaired the Aerial Defence Committee expressed his scorn of those 'enamoured of, I will not say hypnotised by, the blessed word "Science", but while pure science is very well in its way, I think this is a case where it is of more value when diluted by a great deal of practical experience'.[22] Conversely, J. J. Thomson—the world-famous Cambridge professor who had discovered the electron—was frustrated by the attitudes of senior naval officers working with him on a major scientific committee, complaining that the senior staff denigrated it as 'an excrescence rather than a vital part of Admiralty organisation'.[23]

Obeying their patriotic duty, top-level academic scientists temporarily—sometimes reluctantly—took up war-related projects, but, as critics regularly pointed out, that provided no guarantee of success. Ernest Rutherford loyally abandoned his own research and tried to find a way of detecting submarines with sound amplifiers adapted from Marie Curie's laboratory equipment. He was unsuccessful, and it was an independent inventor who thought laterally and suggested sending out sound beams that would be unwittingly reflected back by an enemy ship—in other words, sonar.[24] Similarly, the Army began to lose patience with Nobel Prize winner William Lawrence Bragg as he struggled to create an effective microphone, one that would be sufficiently sensitive to work when the wind was blowing in the wrong direction yet would selectively ignore every sound except enemy artillery fire. Although he did eventually manage to design an effective detector, many other projects by academics fizzled out with no result.[25]

The most productive structure that emerged for achieving technological progress was a mixed regime of industrial, government, and scientific cooperation. British aviation drew ahead before and during the War because private companies were investing heavily in development, university departments were organizing research programmes, the Army was financing work on reconnaissance aircraft, and under Churchill's guidance the Royal Navy was equipping itself with fighters to complement its modern oil-fuelled battleships. Before the War started, a Cambridge professor of mechanical engineering was head of naval education, and Britain had more aeroplanes per military man than any country in Europe. During the War, aircraft production expanded massively, and the government hired senior management experts to coordinate military initiatives. By 1918, British monthly production had rocketed from ten to approaching 3,000.[26]

Radio provides another excellent example of how private companies and state organizations cooperated to ensure scientific progress. Two months before making any public announcement, the Italian electrical engineer Guglielmo Marconi approached the British navy to demonstrate his invention that promised immediate communication over long distances. Immediately recognizing radio's potential for enabling ships to be controlled centrally from London, the Admiralty signed an exclusive agreement with Marconi in 1907. When the War started, they took over his factories, while his staff served on ships and long-range stations dotted along Britain's coastline and around the world. Working together, the Royal Flying Corps and the Marconi Experimental Laboratory developed new equipment—and the pilots were taught how to use it by Marconi engineers.[27]

It served everybody's interests to be seen to be collaborating. Politicians enhanced their reputation by publicly embracing modernity, City investors made small fortunes by trading in scarce scientific commodities, and scientists with expertise to sell manoeuvred themselves into stronger financial and political positions. For example, as First Lord of the Admiralty, Churchill was eager to impress the government by solving the urgent problem of manufacturing vast amounts of explosive. One vital ingredient was acetone, but the wartime restrictions on importing raw materials made it impossible to produce sufficiently large quantities using the same processes as before. When a Manchester University biochemist, Chaim Weizmann, came up with a possible solution, Churchill provided him with a disused gin distillery in East London for scaling up his laboratory procedures to industrial levels. Once the initial trials had proved successful, Weizmann secured the approval of Lloyd George to set up plants all over England.

While serving the nation, Weizmann also benefited personally. Later famous as the first President of Israel, in 1914 Weizmann perceived himself to be marginalized as an immigrant Russian Jew. Feeling that he had been consigned to a provincial university and denied the promotion he deserved, he also saw that his political ambitions were being thwarted—despite years of activity, he still held no official position in the British Zionist movement. By the end of the War, Weizmann had moved to London, struck a lucrative financial deal for his chemical processes, and been put in charge of wartime research at the Lister Institute. Tirelessly campaigning for Zionism, Weizmann used his scientific contributions to steer himself into a powerful diplomatic position, which may (or may not— opinions differ) have helped him persuade the Foreign Secretary Arthur Balfour to pledge British support in 1917 for establishing a Jewish settlement in Palestine.[28]

As the War continued, acronyms proliferated to label the countless committees that were being set up, most of which spawned several subcommittees, which in their turn...! The foundation of the DSIR (Department of Scientific and Industrial Research), which existed from 1916 to 1965, sounds like a resounding success story with a simple plot: 'Top scientists at last persuade reactionary politicians that spending money on science will win the War and keep the Great in Great Britain.' As with so many other propaganda pieces, that was not exactly what happened. It served scientists' interests to downplay the extent of government investment and hence exert pressure for more; they sought greater control over coordinating national policy, but they also wanted science to be state-funded. By the time the DSIR was established, Britain was already one of the most advanced scientific and technological countries in the world, and the big spenders continued to be state-run organizations such as the Air Ministry. After prolonged internal wrangles and some gentlemanly compromises, the Royal Society settled for an advisory rather than an executive role. Because the DSIR was headed by a politician, the government remained in charge of policy, although academics now had some influence over how public money was spent on industrial development; on the other hand, because they retained their independence, the government could not force an unpalatable agenda on the universities.[29]

When investment in science, technology, and medicine increased after the War, opportunities were enhanced for both men and women, although at nowhere near equal levels. Of all the countless War memorials that were erected across the country, one of the most inspired was the foundation of the University of Leicester. Previously, many schoolgirls had been prevented from going to university because they were obliged to act as carers and companions for their parents. By specifically catering for local students, the new institution worked around an obstacle hindering female education rather than trying to destroy it. When Leicester University finally opened in 1921 in a former lunatic asylum, there were only nine students—but eight of them were women.[30]

9

TAKING OVER

Women, Science, and Power During the War

The surest way to keep people down is to educate the men and neglect the women. If you educate a man you simply educate an individual, but if you educate a woman, you educate a whole nation.

James Emmanuel Kwegyir Aggrey, Lecture in Ghana, 1920

In the summer of 1914, just over 46 million people lived in Great Britain. Over the next four years, some of them would become famous, others would permanently switch career, and just about all of them would be affected by privation and fear as well as by death among family, friends, or acquaintances. It is hard to recapture the sense of turmoil, the rapidity with which familiar normality could be abruptly terminated, but as just a couple of specific examples, consider two women who had both been in the process of writing a book when war was declared in August. The Scottish geologist Maria Gordon and the Newnham-trained archaeologist Agnes Conway suddenly found their lives disrupted, but in very different ways.

In 1894, Gordon had become the first woman in Britain to gain a science doctorate. After an extraordinary few years spent defying all the proprieties by exploring on her own in the Dolomites, she married and had children. Still more unusually, she continued to go on field trips, taking her family with her. With many prestigious articles to her name, she decided to write a book-length account of her research, illustrated by her own detailed geological maps. Shortly before the First World War, she handed everything over to a German publisher and went back to Britain for what she intended to be a visit—but found herself stranded there. Temporarily putting geology on one side, she dedicated herself to war work and, as President of the National Council of Women, campaigned for women's rights. The manuscript vanished during the upheaval in Germany, and she spent the next nine years reconstructing it.[1]

Conway's experience was equally unforeseen. A privileged young woman wondering about her future, in August 1914 she was tranquilly writing a book describing her recent tour of the Balkans. Within a week of war being declared, she was helping to equip a hospital, and she spent much of the next two years caring for wounded Belgian soldiers. She also underwent a series of painful but unsuccessful operations to restore her facial nerves and muscles, damaged after falling through a glass roof as a child. On top of all that, by 1917 she had learnt Russian to keep her mind alert, seen her book published, set up a job-finding centre for female graduates, organized a Red Cross depot, and raised funds for charity by selling waste paper.

Her distinguished father was far less busy. A Cambridge professor of art buttressed by inherited wealth, Martin Conway muddled through the War in a state of gloomy depression. But he did take at least one productive step: he recruited his daughter to work for the War Museum, which later opened under his nominal directorship (with the extra label of 'Imperial'). Appointed to a committee that was gathering material about women, Agnes Conway soon amassed a vast collection of photographs, documents, plaster models, and uniforms, and began to organize an exhibition. During October 1918, the Queen and 82,000 other people visited this display of Women's War Work—and a month later, on the day after the Armistice, Conway noted in her diary that 'Peace will mean much more work for me, as the War Museum will be really justified and I want my part of it to be just as perfect as possible.'[2]

Agnes Conway would become better known as an archaeologist, but she was responsible for collating many of the scanty records that survive of scientific women during the War. In 1919, she circulated a questionnaire soliciting information about women's contributions to industrial and scientific warfare, and the replies have been preserved in the archives of the Imperial War Museum, hidden away among 29,068 items about women's work. They provide a fascinating if patchy glimpse into a neglected arena of battle. Often unnamed, female science graduates were becoming food inspectors, Admiralty chemists, public analysts, welfare superintendents for munitions workers, glass technologists, and much more besides.[3] There must also have been many scientific women whose deeds remain unrecorded either there or anywhere else.

The participation of all these women was only rarely acknowledged in public, as emphasized by a chemist at St Andrews University with a charitable heart but an illegible signature: 'I mention these facts as an index of public spirit with which these women gave their services, services which have not received any public recognition.'[4] A century later, they have still been scarcely mentioned, and the long-buried evidence is hard to retrieve.

Life Changes

As the number of employable men in the country diminished, women proved essential for taking over traditionally male jobs. Their experiences varied, but were potentially life-changing. Dorothea Hoffert had studied chemistry at Girton, and she taught at a girls' school until she was requisitioned for research into varnish and food. Now equipped with a high standard of expertise, she was able to enter academia—but like so many women, when she got married she was obliged to abandon her professional career.[5] Other women refused to be deterred by marriage protocols. Elizabeth Williams, a policeman's daughter and scholarship girl, was only nineteen and fresh from her mathematics course at Bedford College when she was recruited to replace a teacher at Nottingham School for Boys. A few years later, she was so angry at being forced to retire when she acquired a husband that the couple set up a small private school, with her installed as headmistress. In her subsequent university career, which included a spell in Ghana and lasted into her eighties, Williams played a key role in reforming teacher training both in Britain and abroad, with a particular focus on improving how mathematics was introduced to primary school children.[6]

During the War, female scientists were able to move into many positions that had previously been inaccessible. As well as being needed to teach, they were called on to inspect factories, survey forests, formulate weather predictions, analyse statistical data, run museums, and carry out war-related chemical and pharmaceutical research. May Leslie, a coalminer's daughter, had received a scholarship to study at the University of Leeds, where she graduated with a first-class degree in 1908. Next she went to Paris, where she worked for a couple of years with Marie Curie on radioactive materials (which may explain why she was under fifty when she died of lung cancer). As one of Ernest Rutherford's protégés, Leslie was hired for secret wartime research into explosives at government factories. Unusually, she was promoted to Chemist in Charge of Laboratory, but—as happened so frequently—lost her position to a man at the end of the War; she did, however, go on to pursue a prestigious academic career.[7]

From the other end of the social scale, Rachael Parsons was educated at Roedean before studying mechanical science at Newnham. When her brother was sent to the front, she took over their distinguished father's large factory supplying turbines to the Navy.[8] Another woman who entered the male world of engineering was Hilda Hudson, who had won first-class marks in mathematics at Newnham. Faced with limited employment opportunities, she reluctantly took up teaching at a Technical Institute for a few years and was relieved to escape by taking a wartime post as a civil servant in the Ministry of Munitions, where she joined a team designing aircraft. In an unpublished letter, Hudson later explained that an initial core of three women had soon grown to twelve. They could already cope with the main part of the work, which 'involved very complicated applications of the first principles of theoretical mechanics', but, thanks to 'the great kindness of the men engineers of the Section', they were also able to learn more complex engineering. 'It was necessary to have an extremely clear and firm grasp of fundamentals,' she reported, 'so as to be able to apply them to an endless variety of conditions.' By the end of 1917, the women were sufficiently expert to carry out calculations on aeroplane strength. Three years after this concealed contribution to the war effort, Hudson retired from salaried work—presumably living off family money—but led a distinguished career as a pure mathematician.[9]

For some scientific women, the War seems to have represented a career hiccup rather than a life-altering event. Ada Hitchins did not get married until she was in her mid-50s, and has left relatively substantial traces of her scientific career, although her early experiences may well have resembled those of contemporaries who have now vanished from the records. Like many of her non-Oxbridge peers, she came from an undistinguished, non-academic background—her father worked in customs and excise. After graduating with a first-class degree from the University of Glasgow in 1913, she won a scholarship to Aberdeen, where she began studying radioactivity with Frederick Soddy, one of Rutherford's former collaborators. Hitchins worked with Soddy on isotopes (a word coined by a female doctor, Margaret Todd). When his male research student was drafted, second-class citizen Hitchins was ordered to leave her own project and take over his. She soon moved into war work herself, leaving Soddy's laboratory to analyse steels for the Admiralty, but when the men came back in 1918, she found herself once again surplus to requirements. She eventually became Soddy's research assistant at Oxford, and when she emigrated with her family in 1927, Soddy provided a glowing reference that enabled her to work in Kenya until 1946 as a geological assayer.[10] At least she fared better than Janette Gilchrist Dunlop, who began investigating X-ray scattering at Edinburgh with the Nobel Prize winner Charles Barkla—her research career was abruptly terminated when she was obliged to become a science teacher because of the national shortage.[11]

In contrast, although Phyllis McKie was only a couple of years younger, the War catapulted her into a professional career that might otherwise have eluded her: because she graduated during rather than before the War, she could move immediately into a military project. The daughter of a Welsh mining administrator, McKie studied chemistry at the University College of Wales, Bangor, and then found a position preparing explosives. She proved so successful at initiating new lines of enquiry that she moved into academia, first in research, then climbing up the academic ladder from demonstrator to lecturer and finally to principal. A similar route was followed by Ruth King, a London graduate of 1914. As a wartime research worker for the Ministry of Munitions, she was awarded an MSc for her studies of picric acid (an explosive), and then became a lecturer at the University of Exeter until 1955. However, as frequently happened to women, McKie and King were so over-whelmed with teaching responsibilities that maintaining a top-level research career became effectively impossible.[12]

Some career scientists were able to combine their academic and their wartime work. Manchester had appointed its first science lec-turer, Marie Stopes, in 1904; shortly after that she became the young-est woman to gain a DSc, and she was also awarded a PhD in Munich. Stopes soon established an international reputation for her work on plant fossils, many of them gathered from the nearby coalfields of Lancashire and Yorkshire. Unconventional and outspoken, she insisted on going down into the mines herself, a gesture that impressed the pit-owners and secured their help. In addition to these local investigations, she carried out research projects all over the world—Japan, America, Canada, Sweden.

By the time of the War, Stopes had already started thinking about family planning, the field in which she would become world-famous, but she was still employed as a lecturer at University College London. As her contribution to the war effort, she joined a new laboratory in Cumberland, which had been jointly set up by the government and Sheffield University to find ways of increasing coal production and reducing explosions in mines. Boasting about her achievements to her mother, she invited her to study reports of a recent conference:

'notice the prominence given to me on the coal question: big-buggy professors who took part & were scarcely mentioned. They are, of course, rightly annoyed.'[13] Her pride was justified: the Director of the Geological Society announced rather pompously that he had been 'astonished at the quality and quantity of the detail given. I went to encourage a young girl, and I remained to learn from a master.'[14]

Some privileged women seized the openings offered by the War as an opportunity to pursue a masculine career that their families would otherwise not have countenanced. Victoria Drummond came from such a high-class background that the Queen had been her godmother—hence her Christian name. On her twenty-first birthday in 1915, Drummond's presents included several extravagances—a silver-blue fox fur boa and muff, a silver-mounted umbrella, two sparkly shoe buckles from Paris—and also a very special gift from her father: her choice of career. At any other time, he might well have rejected her choice of marine engineering, but a year later, she began by taking up a wartime apprenticeship at a local garage. Clad in her newly purchased overalls, and strongly encouraged by her manager, a former naval engineer, she spent the next two years cleaning gearboxes, dismantling engines, and studying maths at home in the evenings.

Drummond's life changed again in 1917, when the friendly manager was sacked for drunkenness and she started training at a nearby ship works. Her prestigious family connections helped her to get a start, but she had to earn her high reputation for herself. In 1941, she was still the only woman to be fully qualified as a ship's engineer, and she was awarded an MBE for her bravery and cool-headedness when her ship was bombed. As the official report put it,

> When the alarm was sounded, Miss Drummond at once went below and took charge. . . . She nursed this vital pipe through the explosion of each salvo, easing down when the noise of the aircraft told her that bombs were about to fall, and afterwards increasing steam. Her conduct was an inspiration to the ship's company, and her devotion to duty prevented more serious damage to the vessel.

Hard to believe that, as a small child, she had crept down from her nursery and handed home-made toffee to the Prince of Wales.[15]

Other women also benefited from wartime openings in engineering. An archival letter describes how some 'well-educated women' received on-the-job training at the British Westinghouse Electrical & Manufacturing Company, which initially employed them as assistants for calculating graphs, but soon decided that male staff should educate them still further with eight hours of lectures a week. The women gained by being able to do more challenging jobs, and the company benefited by employing skilled labour at a lower cost than men with six or seven years of training. Apart from those who left for 'personal reasons' (a euphemism for 'getting married'?), these women stayed with the company after the War.[16]

Going into engineering via a university degree course seems to have been more unusual than such hands-on routes. The first woman in the world to graduate in engineering was Alice Perry (Galway, 1906), but the War made it a slightly more feasible choice and applications rose. The first Scottish woman to complete her course was Elizabeth Georgeson, who was twenty-one when she entered Edinburgh University in 1916: possibly (but this can only be surmise) she had waited until reaching the age of majority before defying her parents to take this unusual step. Frustratingly, no information survives about her subsequent career. Perhaps like so many young scientists, she was forced to resign when she got married; or perhaps she retreated into a more conventional job, defeated by resistance to employing women in this exclusively male field.[17]

As a very different type of industrial example, Margaret McKillop was fifty when the War started. A schoolmaster's daughter and an Oxford–Dublin mailboat graduate, during her long teaching career she repeatedly campaigned for more women to enter science at a professional level. At King's College London she set up courses in domestic science, seeking to put home economy on a respectable academic footing paralleling agriculture, medicine, and engineering. During the War, she worked at the Ministry of Food (gaining an

MBE) and became intrigued by the new techniques of management efficiency being developed in America. Together with her son, in 1917 she published a pioneering book, *Efficiency Methods*, which was effectively an illustrated instruction manual showing how to make industry less wasteful and more productive. Although she antagonized the burgeoning trade union movement, the War put McKillop in a strong position for applying emotional pressure to adopt her recommendations: 'In a time when everyone's work, when real work, has the burden and privilege of being in its small or great degree national,' she wrote with patriotic zeal, 'to be inefficient has become a moral crime.'[18]

The experience of Margaret Turner, a chemist responsible for preparing large-scale supplies of essential drugs at Aberystwyth, was probably more typical. Although barely mentioned in reports, she wrote a heartfelt letter to the Royal Society's War Committee before it was disbanded in 1915:

> I was one of the workers in the preparation of diethylamine some weeks ago and should be glad to hear of any further help I could give. I can put all my time and energy at your service for the next 6 weeks, and am anxious to know whether the few helpers down here, could not be allowed to contribute further to the needs of the country?

Women's contributions became less crucial after chemistry was put on the reserved occupations list: the reference for one job applicant emphasized that he was a '*man* and a chemist'. Surviving evidence suggests that some of these male scientists were opportunistic rather than patriotic—almost a third of those who asked for exemption were turned down, and even one who *was* allowed to stay refused to accept a post he was offered because it paid only £250 a year. Compare that with Turner's attitude of self-sacrifice for the national cause: 'I, for one, am willing and eager to give up all ideas of holiday while there remains so much to be done.'[19]

Younger women studying science also became involved, presumably often in their enthusiasm defying parents and teachers. In 1917, a Newcastle college magazine reported that 'The latest to go are Miss Oliver, Miss Clayburn and Miss Cummings, late candidates for Honours in Chemistry, now chemists to munition works at Middlesboro.' Whether or not those three eventually returned to complete their degrees, they must have been joined by patriotic volunteers from other universities. Some schoolgirls did at least start their war work with the intention of resuming their studies later: 'Gwendolen A. James is working in an analytical laboratory for a year before going on with her Science Degree work', reported a school magazine; 'Margarethe Mautner is analysing drugs in a Chemical Laboratory before taking up her medical studies.'[20]

Apparently no records survive to indicate the careers of Gwendolen and Margarethe, but like many other schoolgirls of the time, they may well have returned to science. In 1915, Sheina Marshall took a year out from her studies to paint luminous instrument dials in her uncle's factory, but she went on to gain a first-class degree and become a distinguished marine biologist. After the War, she benefited from the government's increased investment in science to secure a permanent post in a Scottish marine laboratory studying food chains, and she became the eighteenth female fellow of the Royal Society. And like many other scientists and doctors of her generation, in the Second World War Marshall interrupted her career once again, this time to investigate how seaweed could be used.[21]

Changing Places

Museums provided an important if equivocal opportunity for women—even those without a degree—to follow a scientific career. When conscription was introduced at the beginning of 1916, staff levels plummeted, and the government decided to close down all the museums, both as an economy measure and to reinforce its austerity regime. Public outrage was loud, immediate, and fanned by the press. The museums are the universities of the people, protested crowd-stirring journalists. Closing them would save a mere £50,000 a year, objected the *Evening News*—which would 'pay for the war for *very nearly quarter of an hour!*' (original emphasis).[22] The First World War was indeed enormously expensive; the full debt was only finally paid in early 2015, at a time of recession when stringent official economy measures were once again causing widespread hardship, but a dramatic political gesture seemed worthwhile.

While the War lasted, the Natural History Museum in London also functioned as a military education centre, giving advice on how to deal with a startling variety of questions: How do you treat a man with a leech up his nose? What is the best sort of wood for building an aeroplane? How can you identify seeds that are poisonous for horses? In order to learn about camouflage, soldiers visited the Museum's displays of animals' protective colouring, while civilians were shown grisly models of houseflies and putrefying food as part of the practical advice on health and hygiene.

From the beginning of the century, several museums had employed women in low-paid jobs as assistants, cataloguers, and illustrators. Annie Lorrain Smith, for example, had three brothers who became university professors, but she never even got a degree. She did, however, become a great expert on lichens, seaweed, and fungi, and the book she wrote in 1921 was still the standard reference work fifty years later. She occupied important posts at the Natural History Museum, carried out field research in Ireland, and more or less took over the department during the War. Although she worked there almost all her life and was awarded an OBE for her science, as a woman she was always classed as an 'unofficial worker' and paid (poorly) from a special fund.[23] Similarly, while Helen Fraser was studying zoology at King's College, she worked part-time at the Museum carrying out routine tasks of labelling and preparing diagrams. Even after a new curator reduced her hourly rate of pay, she stayed on, seizing the opportunity to carry out unpaid research into rust fungi.[24]

One museum pioneer in particular stands out from this period—the palaeontologist Dorothea Bate, who had been working at the Natural History Museum since 1898. In 1914, as the first volunteers were leaving for the front, she moved in to fill the gap. Working in dimmed lights to guard against aircraft attacks, she started by cleaning and preparing fossils; as the department shrank, she took on more and more responsibilities, coping with visitors as well as specimens arriving from collectors overseas. But—and this is a big but—for thirty-seven years, she was officially employed only on a temporary, per-hour basis, and her take-home pay was substantially less than that of far less knowledgeable male assistants in salaried positions.

After the Armistice, some museum women became professional research scientists in their own right. The Natural History Museum's workload increased as curators tried to catch up with the backlog of unclassified specimens, and in order to replace members of staff who had been killed, a few more women were employed. But, like Dorothea Bate, they received denigrating treatment. Helen Muir-Wood, for example, was already on her way to becoming an expert on ancient brachiopods (shelled animals living in the sea), but she was routinely addressed as 'Miss' rather than 'Doctor'. The temporary contracts on offer were so mean that no men would apply, and it was not until 1928 that women could apply for permanent posts. Even then, the advertisement made it clear that they would be paid less than men for carrying out the same work.

Other institutions were almost completely turned over to wartime research, and so new—if often temporary—openings appeared for women. For instance, two out of the seven science graduates in the chemical laboratory of Kingsmouth Airship Station were women, where they carried out important research into improving airship fabrics and reducing the effects of weather.[25] At the Lister Institute of Preventive Medicine, forty women—seventeen of them with university qualifications—were employed in work having a direct bearing on the country's wartime activities. Together they published an impressive list of academic articles, but in the absence of trained assistants, they recruited any willing volunteers they could find, such as women to look after the horses used for making anti-toxin sera. Similarly, the illustrator was enrolled to help with diagnostic tests, and the laboratory cleaner, Mrs Soper, stayed late into the night looking after the experimental rabbits and feeding them with milk from a fountain-pen filler. As the team head Harriette Chick put it, 'Many a tommy may have owed his life to Mrs Soper.'[26]

Scientists who graduated during the War were lucky enough to have their first experiences of research in environments that already included female scientists. Constance Elam, for example, went from Newnham to the National Physical Laboratory at Teddington, which had lost around a quarter of its male staff. She started there in the metallurgy department, and the following year went to Imperial College, where the department had shrunk from twenty studentships to four. In the near-absence of young male researchers (there were seven women from Newnham alone), she worked closely with an eminent professor on the effects of heating metals. Given this solid start because of the War, Elam subsequently went on to become a distinguished metallurgist.[27]

The experience of university life was very different during the War. Most obviously, there were relatively few men around because they had either volunteered, been conscripted, or tried to avoid being sent overseas by taking refuge in munitions factories. Especially during the vacations, female students joined the Land Army or carried out some other form of voluntary war work. Although older male lecturers could have stayed behind, many of them enlisted, offering their technical expertise to build railways or sweep mines or develop explosives. Their absence forced departments to appoint women as temporary lecturers, a practice that had previously been virtually unheard of: in these male-dominated institutions, although women might have been tolerated for back-room research, they were generally deemed unsuitable for lecturing to young men.

Inside and outside laboratories, women were surrounded by reminders of the War. Wounded soldiers were housed not only in schools, colleges, and churches, but also in temporary hospitals made from tents affording little protection for the cold nurses tending miserable convalescents. Emergency drills and blackouts became as familiar as the meagre rations of semolina and lentil pottage.[28] At Oxford, the all-female Somerville College was taken over by the War Office and converted into a hospital annex, while the displaced students went to live in lodgings formerly reserved for men or in carefully segregated sections of Oriel College. Apart from the disabled, overseas students, and the Cadet Corps, male undergraduates were absent from the oddly deserted streets.[29]

Life in Cambridge was transformed by the arrival of the army and many thousands of wounded soldiers, who were cared for not only in schools and colleges, but also in a massive military hospital with 1,700 beds that was hastily constructed on a playing field (see Figure 9.1; it is now the site of the University Library). Following fashionable medical theories, the wards were open-sided to permit the circulation of fresh air. The nurses complained bitterly, but they suffered not only from the severe weather but also from the cats and the rats. New patients

Figure 9.1. First Great Eastern Hospital, Cambridge.

were transported in horse-drawn carts through town from the railway station over a mile away, nurses were billeted in nearby colleges and convalescents frequented the shops. The two women's colleges were particularly involved in this military initiative. Girton's secretary was in charge of administration, and female medical students were recruited to compensate for the shortage of doctors: after only a week's gruelling training, one young woman from Newnham found herself administering anaesthetics.[30]

In the absence of many male students, places became available for women on regular university courses. At Manchester, for instance, the percentage of female students rose from 21 to 43, and they switched away from arts subjects to more vocational courses, such as medicine, engineering, and science.[31] In Cambridge, Newnham's Balfour Laboratory was no longer needed: elderly lecturers and women acted as demonstrators in mixed facilities. Some recalcitrant professors refused to accept that their lectures would be packed with women. As one cynical Newnham student remarked, 'the preponderance of women in the classrooms made the salutation of "Gentlemen" more ridiculous than ever.'[32]

The War presented an opportunity for the newer universities to demonstrate their national importance. Drafting in female academics to compensate for the shortage of men, they matched their research programmes to military requirements. Chemists were in particularly short supply. At Sheffield, for example, there was only one man left in the chemistry department, which hired women both to teach and to synthesize large quantities of local anaesthetic needed for the War. Similarly, at the University of Wales, women developed drugs and explosives for the Ministry of Munitions, while in the laboratories at Cheltenham Ladies' College, students were trained to manufacture anaesthetics.[33]

Imperial College London was especially heavily involved in wartime activities. The grounds started to resemble a military encampment, equipped with experimental trenches dug in the gardens, where a largely female team developed an improved grenade filling. Several buildings were taken over by the Army as offices for clerical workers or as accommodation for recent recruits. Reportedly, in the mornings scientists had to step over soldiers sleeping in the corridors, while museum cases were used as dining tables. Making the conditions still more cramped, the college housed various refugee groups from Europe and the British Empire. Twenty women worked in the Royal Flying Corps drawing office, and space also had to be found for the volunteers in the Women's Emergency Corps. As well as being allowed to lecture, female scientists were given high positions in research projects. Miss Lodge and Miss Jackson were promoted to supervise the fly room for developing insecticides, and an eight-woman team tested explosives and poisonous gases. And when a male student accidentally gassed himself with phosgene, it was a woman who rescued him (they later married).[34]

Academia, industry, and government cooperated more than ever before, although stories of mutual incomprehension abound. Several may be apocryphal: can it really be true that a military officer told a scientist the Army had no need for a meteorological service because the British soldier fights in all weathers?[35] Unsurprisingly, many of the scientists involved were men, but the general pattern of collaborative national effort is indicated by the vast range of activities involving women, who were scattered far and wide—developing anti-aircraft systems at Woolwich, improving bacteriology in tropical hospitals, inspecting valves for the RAF, researching optical glass manufacture at the National Physical Laboratory, code-breaking in naval intelligence, testing trench bombs, studying the effects of personnel fatigue, analysing tear gas, vetting insecticides, designing aeroplane equipment. Female scientists were also investigating how they could sun-dry vegetables to retain their vitamins, and studying fungi at the Royal Botanic Garden: in wartime, the need to find out why bread goes mouldy was as pressing as the invention of a new explosive.

Both during and after the War, women were paid substantially less than men for the same work, and many female volunteers worked long overtime hours for nothing. Reporting on women's wartime research into synthetic drugs, explosives, and poisonous gases, a Scottish university professor wrote:

> It should be noted that the whole of this work was unpaid from Government sources, the workers receiving only their University salaries, in cases where they were members of staff, or the value of their Scholarships, if they held any such distinction. Not only so, but the demand for chemists throughout the war was continuous, so that the workers who remained with me gave up many opportunities for professional advancement.[36]

To allay industrial unrest, in 1917 the government introduced a bonus scheme in compensation for the massive amounts of overtime work caused by the War. Unsurprisingly, the awards given to women were only about two-thirds of the men's, although some institutions—Imperial College, for example—decided that such overt discrimination was wrong, and allocated some of their limited funds to equalizing salaries. That sounds like an enlightened move, but they were undermining the only advantage women had—being cheap to employ. Which is better: to have no job or a low-paid job? As many women discovered, there is no straightforward answer to that question.

10

CHEMICAL CAMPAIGNERS

Ida Smedley and Martha Whiteley

It is with much interest that we learned a few weeks ago that women chemists in London had formed a Club. Most men are clubbable, one way or another, but we did not know this was true of women. We wonder if this formation of a Club for women chemists is another sign of female emancipation. We should be glad to think that they mellow over a bottle or two of fine wine. We commend claret—the Queen of wines. Presumably claret obtained this title because of its beauty, its grace and its subtlety—admirable qualities which men have always associated with women.

Editorial, *Chemistry and Industry*, 23 February 1952

Hostility towards women in uniform could be quite extreme. As a former munitions worker, Olive Taylor knew all about the mockery endured by yellow-skinned canaries. But after she joined the Army, the abuse got worse. Only once did she dare venture from her military camp into the nearby town with her friends. Despite being patriotic supporters of the war effort, they 'were subject to such insults that we never went again', she protested; 'we learned for the first time that we were regarded as the scum and that we had been enlisted for the sexual satisfaction of the soldiers.'[1]

Female scientists were confronted with expressions of prejudice that were usually more subtle but could be just as hard to deal with, especially when lying concealed beneath a surface veneer of friendliness. Consider nicknames, which can be simultaneously cruel and affectionate. Inside the Admiralty, wartime cryptographers thrust together by the need for secrecy assigned each other characters from

Alice in Wonderland—the Dormouse, the White Rabbit—a game that the collusive participants presumably enjoyed. But how did their boss feel about being known publicly as 'Blinker' Hall because he had a nervous tic in his eyes? And what about the female filing clerks—did they truly relish being referred to as the 'Angels of room 47'?[2] After the War, nobody in Sheffield's chemistry department admitted to remembering why Dorothy Bennett and Emily Turner, both university trained with MScs in chemistry, had been dubbed 'the Tartrate Twins'.[3] Looking back, the reason seems clear: recruited as assistant lecturers and demonstrators in the absence of men, these two women in a man's world were bracketed together as intruders. It was precisely such discriminatory attitudes that two campaigning chemists, Ida Smedley and Martha Whiteley, were determined to stamp out.

In universities, industrial laboratories, museums, and research centres, female scientists encountered resentment and rejection on a regular basis. Why, wondered an Edinburgh student, did traditional male chivalry vanish as soon as she and her friends walked into a classroom? 'The tapers and tadpoles of the back benches begin to howl and screech with all the lustiness of rural louts,' she remarked bitterly, 'and seem to be as much amused as if they saw a picked company of the far-famed Dahomeyan Amazons march in all the glory of their military attires.'[4] Cambridge's young women encountered similar performances: 'in the lecture theatres, the men stamped their feet while the female students went to sit in the reserved rows at the front.'[5] Superficially, male lecturers behaved more politely to female colleagues who were technically their equals, but the effect was the same: 'as a woman I had no place at all in [Bristol University's] government. I was a member of no Faculty or Board of Studies. I was in a sense an outsider.'[6]

Operating more deviously, the assistants in the mixed laboratories at Cambridge deliberately neglected to lay out all the equipment at the female students' places, thus forcing them to walk up to the counter and wait while everybody stared. Newnham had its own laboratory for the life sciences, but that did not mean women benefited from the same resources as the men. 'All equipment had to be bought or borrowed, the strictest economy having to be exercised in every way,' complained the geneticist Edith Saunders. 'Reagents had to be purchased, apparatus set, microscopic sections prepared, with only a young untrained boy to act as general laboratory attendant.'[7]

On a daily basis, these stark messages that women did not belong in the academic world of science were reinforced by inferior facilities for relaxing and socializing. Between lectures at Bristol, the first women had only one place to go: 'the small women's cloak room. It was furnished with three or four wooden chairs and a small deal table and with pegs for our heavy cloth waterproofs. Here we worked and ate the lunch we had brought with us.'[8] By the time of the War, the redbrick universities were generally supplying proper common rooms for both male and female students, but at Birmingham only one of them (you can work out which) resembled 'any other first-class club, and possesses a dining-room, smoke rooms, reading rooms, billiard rooms, etc.'[9] When Imperial College decided to spruce up its dingy women's common room, wives of the older scientists did the decorating, and the rector's wife donated a handsome silver tea service. Problem: how was the water to be heated for the tea? Solution: on a single gas ring inherited from an older laboratory and located in the ladies' toilets.[10]

What motivated these pioneering female scientists to persist against such obstacles? Vague phrases such as 'the love of science' blandly skim over daily realities, but perhaps some women genuinely did share H. G. Wells's idealized vision of a biology laboratory:

> The whole place and everything in it aimed at one thing—to illustrate, to elaborate, to criticise and illuminate, and make ever plainer and plainer the significance of animal and vegetable structure. It dealt from floor to ceiling and end to end with the theory of the forms of life; the very duster by the blackboard was there to do its share in that work, the very washers in the taps; the room was more simply concentrated in aim even than a church.

Perhaps for disillusioned and weary campaigners, research did indeed provide a welcome refuge from the confused movement and presences of a Fabian meeting, or the inexplicable enthusiasm behind the suffrage demand, with the speeches that were partly egotistical displays, partly artful manoeuvres, and partly incoherent cries for unsoundly formulated ends, compared with the comings and goings of audiences and supporters that were like the eddy-driven drift of paper in the street, this long, quiet, methodical chamber shone like a star seen through clouds.[11]

As Britain's two most prominent chemical campaigners for female equality in science, Smedley and Whiteley are often twinned together. Both pursued outstanding but very different scientific careers while also endeavouring to improve conditions for women, and after a protracted battle, they were among the first group of women to be allowed membership of the London Chemical Society. Between them, Smedley and Whiteley did so much to change the lives of so many female scientists, both then and in the future. But apart from their scientific articles, they have left only fleeting traces of their existence. Their short entries in the *New Oxford Dictionary of National Biography* sketch out the bare facts of their lives, although frustratingly little information survives about their feelings and experiences—but a substantial amount can be inferred.

Ida Smedley (1877–1944)

Ida Smedley (Figure 10.1) grew up in an extraordinary family. Her grandparents had lost their trading fortune when the plantation slaves in the West Indies were freed, and so they taught their daughter to earn her own living—and she in her turn made sure that her two daughters, Ida and Constance, were exceptionally well educated and independent. Originally an accountant, their father became a wealthy businessman and was an eminent local philanthropist: the two sisters were recruited for the Ladies' Street Collectors, standing with their mother outside the railway station to raise money for charity. With only a year between them, as children they played music and acted together, shared their father's love of Shakespeare, and conversed on

Figure 10.1. Ida Smedley in 1933.

equal terms with the distinguished visitors who came to the regular mini-salons at their Birmingham home.[12]

Whereas Constance, despite being severely disabled, went to Art School and pursued a distinguished literary and theatrical career, Ida chose to become a chemist. Both girls had attended King Edward VI Grammar School, the best girls' schools for science in the country. Along with a few other favoured pupils, Ida was allowed to join some men's classes in physiology at a nearby college, overcoming the objection that 'it was indecent for a girl to study biology'. One of her schoolmates described how they entered 'the sacred medical department of the College, and one of the demonstrators…announced that he would leave if those little girls from the High School came to his laboratory. He did not keep his word.'[13]

At school, the Smedley sisters benefited from first-class teaching by Cambridge graduates. The headmistress, Miss Creak, had been one of the original five Newnham undergraduates, and women moved between the school and the college in both directions—as students leaving and as qualified teachers returning. So when Ida finished school in 1896, it seemed natural that she would go to Newnham. Because the College had its own laboratories, she had no need of a chaperon for mixed classes, but like the other students, she would have worn

> almost a uniform of white blouse, and tweed coat and skirt. The blouse should be of 'nun's veiling' and the collar fixed with a plain rolled gold pin; but such blouses were expensive!... not on the warmest spring day might that coat be removed, either at lectures or in the street.[14]

After two years at Newnham, Smedley won first-class honours in part one of the Cambridge sciences tripos, and the following year was awarded a second-class in part two. After graduating, she managed to avoid the most usual destination of school-teaching. First she won a Bathurst scholarship, which enabled her to carry out research for two years under the notoriously misogynistic Henry Armstrong. Then followed two years back at Newnham, and after that, another spell in London as a research fellow at the Royal Institution. In 1905 she was awarded a London DSc (Doctor of Science), which was a very advanced research degree; it was not until 1917 that British universities began introducing lower-level PhDs (Doctor of Philosophy), under encouragement by the government to prevent research students leaving for Germany or the USA.

The following year, Smedley was off to Manchester, the first woman to be appointed a member of staff in their chemistry department.[15] For four years, she taught there as an assistant lecturer, at the same time carrying out original research into the optical properties of organic compounds. An essay written by a student who made a similar transition indicates what Smedley might have felt about moving to a redbrick university. At Cambridge, remarked 'A Manchester Girl', 'it is easy to forget the world, to dream one's dreams, to see life as a pleasant wandering between the library and the study.' In Manchester, however,

> Life in its starkest aspect thrusts itself upon her as she threads her daily way through the hurrying crowds of her city-home. As regards scientific

appraisement and impartial judgement of problems and values, the Cambridge student has the advantage.... But life has to be lived before it can be understood, and the Manchester girl sees things from the centre.[16]

Smedley presumably also shared some experiences with the paleobotanist Marie Stopes, Manchester's first female lecturer, who had arrived a couple of years earlier. In contrast with her time in London, Stopes felt welcomed by immediately being made a member of the Professors' Common Room. However, although—like Smedley—she had a stellar academic track record, she felt under constant close scrutiny from older faculty members and their wives. Since Stopes established a satirical botanical journal called *Sportophyte*, it is difficult to know how seriously to take her reports of academic life, but apparently the botany professor warned her bluntly that she was a risky experiment. According to her, she learnt how not to lecture by attending classes given by older men. One eminent professor was apparently taken aback by the roars of laughter that greeted him when he started a lecture by drawing a wiggly shape on the blackboard and announcing: 'Let us suppose that a circle is a circle.'[17]

In 1910, Smedley scored another female first by being awarded a Beit Memorial Research Fellowship. This exciting achievement was announced not only in her Birmingham school's newsletter but also in Australia, where the *Western Mail* singled out Smedley for special mention in an article, 'Women as Inventors', which feted other notable celebrities such as Marie Curie, Hertha Ayrton, and Marie Stopes (although to modern eyes, this feminist message was somewhat undermined by an adjacent item advising suitors how to analyse a girl's character scientifically from her eyebrows).[18] Smedley was only the fifth recipient of a Beit award, but up to 1936, nearly 15 per cent went to women, a surprisingly high proportion.[19] Equipped for the time being with her own funding, Smedley moved on to the next stage in her prize-studded career—London's Lister Institute of Preventative Medicine, where she worked on the metabolism of fats and won an American prize for outstanding research by a woman scientist.

Unlike many scientific women of this period, Smedley continued her career after she married Hugh MacLean, a fellow researcher who later became a professor of medicine (their romance developed on the golf course as well as in the laboratory). Along with many of her male colleagues, Smedley temporarily put her own research to one side during the War to work for the Admiralty. In particular, she participated in the project for which Chaim Weizmann got the credit—the industrial-scale production of acetone, which was needed for making explosives. Somehow she managed to cope with looking after two children as well as her sick husband and disabled sister, working continuously until she died in 1944. Was there perhaps enough family money to pay for full-time domestic staff? By then, she had published around thirty papers as well as an important book on her major topic, the metabolism of fat.

Throughout their adult lives, both Ida and Constance Smedley were deeply involved in feminist action. Although they shone in very different spheres, they had been so close as children that they continued to support each other. An artist and playwright, Constance was known by two nicknames: 'Peter' (the miller's third son), to represent her longing for masculine freedom and adventure; and 'Princess', perhaps ironic but supposedly to indicate her benevolence and imagination (one of her plays, *The April Princess*, was a self-referential dramatization of a daughter's rebellion against her philistine Victorian family). Ida played the leading role in one of Constance's early plays about female independence, and later represented 'the Universities and Science' on a committee for one of Constance's major projects—establishing the International Lyceum Clubs to support professional women when they travelled.[20]

A champion of the suffragettes and at one time a newspaper reporter, Constance braved attacks from hostile men by following processions to comment on protesters who were chaining themselves to railings. Perhaps after consultation with her sister Ida, she wrote an imaginary letter urging a stereotyped science professor to stop obstructing women who were carrying out the important business of researching into life:

> [L]eave those women who are thinking deeply, striving highly, working earnestly, to experiment and work out their development.... [I]f their energies are being wasted in abortive directions, surely you, dear Herr Professor, should be the last person to be severe on those who are adventuring their lives in the pursuit of what may be fruitless, but which at least are conscientious experiments!

This formed part of a small book Constance published, a confrontational riposte to entrenched views on women's behaviour: witty but provocative, it aroused objections ('brickbats', she called them) from women as well as from men.[21]

During Ida's first year at Newnham, when an attempt to award women degrees proved unsuccessful, male students celebrated by rioting in the streets and hanging female effigies. Whatever her reactions then, she became increasingly involved in feminist politics, and in March 1907 convened a meeting of seventeen academic women in a Manchester girls' school. Angry about the obstacles facing women, she 'held forth vigorously as we sat on the benches and she advocated the formation of a Federation of University Women with the objects of helping women to promotion on university or similar staffs and raising money for research'. A couple of months later, her vision had been realized. The British Federation of University Women (BFUW) was formally founded, and before it moved to London a few years later, Manchester acted as its hub for a range of initiatives throughout the country. After Smedley went to the Lister Institute, she continued to play a key role in the Federation, working closely with the physiologist Winifred Cullis, a friend from both school and Newnham. Based at the London School of Medicine, Cullis later became Britain's first female science professor. Together, Smedley and Cullis ensured that the BFUW was closely linked with women's medical organizations. From 1929 to 1935, Smedley (now MacLean) was president of the BFUW, which still offers a research fellowship in her name.[22]

Smedley also played a prominent role in the protracted battles at the Chemical Society...and that is where she joined forces with another exceptional chemist, Martha Whiteley.

Martha Whiteley (1866–1956)

Eleven years older than Ida Smedley, Martha Whiteley (Figure 10.2) came from a less privileged background, yet their paths became intertwined through their shared commitment to science and to women's rights.[23] Whiteley's father was a surveyor, and she went to one of the high-powered London high schools financed by the Girls' Public Day School Company. When she went to Royal Holloway in 1887, at that time a residential women's college associated with the University of Oxford, she was in the first intake of twenty-eight students, her fees of £90 more or less covered by prizes and scholarships.[24] The chemistry laboratories had not yet opened, so the four science students used three rooms equipped with sinks and taps, 'and

Figure 10.2. Martha Whiteley at Imperial College.

did Chemistry, Physics or Botany according to the way the wind blew, because one chimney always smoked'. The daily regimen was strict: a wake-up bell at 7 a.m., a compulsory service at 7.55, and breakfast at 8.20 before lectures from 9.00 to 1.00. The early afternoon was given over to tennis and other games, followed by tea, private study, dinner, and evening prayers.[25]

Whiteley graduated with a University of London degree when she was twenty-four, but, lacking either rich parents or a husband to support her, she spent the next eleven years teaching. Although no hard evidence survives of her 'Dear Diary' feelings about following this route, her frustration is suggested by the fact that for several years she was carrying out scientific research as well as working to earn her living. The year after getting her DSc in 1902, she joined the staff at one of the three colleges that soon merged to form Imperial College. With the subsequent help of a fellowship from the BFUW that Smedley had helped to found, after several promotions at Imperial she eventually became the equivalent of a modern reader until she retired in 1934; she was made an honorary fellow in 1945.

Whiteley was clearly an excellent teacher: a student recalled that, just after the War, she 'was in charge of the undergraduates' organic laboratory.... They were nearly all men and a very lively lot—being mainly ex-Service men, but she had them completely under control. She managed to turn them into fairly tidy, efficient practical workers.'[26] In collaboration with her former supervisor, Jocelyn Thorpe, to whom she paid great tribute for his support, she wrote a four-volume encyclopaedia of organic chemistry. Based on her long experience of running the advanced course, it became a standard text and over the years was produced in several editions.

Whiteley's distinguished career was unique for a female chemist in this period. Several women found places as assistants to men, and a few—notably Ida Smedley—became renowned for their research in a non-traditional field. For a woman to stake out her own patch in unexplored territory was the quickest, if not the surest, way to an outstanding reputation. To take the most obvious example, Marie

Curie was one of the very first scientists who investigated the new phenomenon of radioactivity. Constantly worried about her claim to ownership in a country where as a wife she was legally a non-person, Curie repeatedly and systematically buttressed her assertions of priority.[27] But Whiteley was different: she chose a well-established topic (amides and oximes, types of organic molecule), set up her own research programme with both male and female students, and worked independently for many years at a major educational institution, Imperial College.

In books about the War, references to Whiteley (those few that exist, that is) are as terse as the note she asked her secretary to type and which, thanks to Agnes Conway, is now in the archives of London's Imperial War Museum. Bound by military secrecy, in 1919 Whiteley answered Conway's enquiry about women's war work by reporting only that research had been 'actively prosecuted in the Organic chemistry laboratories' to prepare 'certain substances for the Naval hospitals' and trench warfare. Whiteley headed a team of seven women, five of them working, like her, on a voluntary basis—Dorothy Haynes, Winfred Hurst, Hilda Judd, Frances Micklethwait, and Sybil Widdows. All proficient research scientists, they have been virtually forgotten, as have the two graduates they employed, Louise Woll and Ethel Thomas.[28]

When Thorpe left London to investigate toxic chemicals in France and at Porton, he put Whiteley in charge of the experimental trenches and the temporary workshop installed just outside the main chemistry laboratory at Imperial College. Putting on one side her research into synthesizing barbiturates and other drugs (presumably those 'certain substances for the Naval hospitals'), Whiteley shifted to examining gases. And there was only one way for her group to do that effectively: by testing the gases on themselves. Although they did not share the fate of other wartime chemists, who died through such self-experimentation, they went through some unpleasant experiences. Over thirty years later, in a lecture designed to inspire female students, Whiteley described how she had examined the first sample of mustard gas to be brought back to London. 'I naturally tested this property by applying a tiny smear to my arm and for nearly three months suffered great discomfort from the widespread open wound it caused in the bend of the elbow, and of which I still carry the scar.'[29]

Whiteley received several tributes for her wartime research. She must have felt gratified to have an explosive named after her—DW for Dr Whiteley—and also proud to be awarded an OBE. More unusually, she was celebrated in the press as 'the woman who makes the Germans weep' because of her research into tear gas.[30]

Whiteley had been President of the Imperial College Women's Association since its foundation in 1912, and she also played a prominent role in the protracted battles at the Chemical Society . . . and that is where she joined forces with Ida Smedley.

The Forty Years War at the Chemical Society

Of all the forms of explicit exclusion, perhaps the one that rankled most was being banned from professional societies. By the end of the War, female scientists were running their own research projects, lecturing at universities, writing textbooks, advising industry, inspecting factories—but they were not allowed to join either the Royal Society or several other specialized groups. Although in practical terms the Royal Society's doors remained closed until 1945, women first entered the Chemical Society in 1920, after forty years of petitions, debates, and manoeuvring behind the scenes. The details of who did what when can become very complicated, but this clash between two gendered scientific communities illustrates how women created their own networks and gained strength from their fellow activists. It also, less creditably, shows how much influence a few powerful men could exert.[31]

The Chemical Society was founded in 1841, but it was not until 1880 that a couple of members suggested amending the constitution to allow women to enter. On three separate occasions, that proposal was shelved, but in 1904 Marie Curie was recommended for fellowship. What an awkward situation! Yet again, legal advice was taken, and the final decision was an inspired compromise of the type all too familiar to anyone who has sat on committees: although as a married woman Curie was banned from being admitted normally, she could be a Foreign Fellow, a position that carried no duties or responsibilities.

At this stage, Smedley and Whiteley decided it was time for action. Gathering together seventeen of their female colleagues, they sent in a petition pointing out that during the previous thirty years, around 150 women had been named as authors in the Chemical Society's own publications. The signatories came from across the country, but several of them had connections with Newnham and/or Royal Holloway College, where Smedley and Whiteley had (respectively) been undergraduates. Cue another near-farce: although the Council unanimously adopted their proposal, the Society's 2,700 members were less enthusiastic—only forty-five even bothered to attend the special meeting, and their vote was twenty-three against, twenty-two for.

The story gets funnier—well, at least it would if it weren't so poignant. The president of the Society, whose research group included Whiteley, persuaded 312 distinguished members (such as Whiteley's supervisor, Jocelyn Thorpe) to sign their own petition in favour of admitting women. In *Nature*, editorials poured derision on the recalcitrant chemical cabal, and this time almost two-thirds of the voters approved the recommendation. All done and dusted, it would seem. But no! One of the most belligerent opponents was Smedley's own research supervisor, Henry Armstrong, who maintained that the prime duty of female chemists was to pass on their genius to the next generation by producing baby chemists. Young women should, he insisted, 'be withdrawn from the temptation to become absorbed in the work, for fear of sacrificing their womanhood; they are those who should be regarded as chosen people, as destined to be the mothers of future chemists of ability'.[32] At the Council meeting, Armstrong—a former President and a domineering man who served on the Council for over sixty years—convinced two-thirds of those present to reject the Society's own referendum, and instead to give women subscriber status, which implied partial membership with limited rights. Only eleven women ever accepted that grudging invitation.

The following year, in 1909, someone (Armstrong?) circulated a letter accusing the women agitating for membership of being linked to the suffrage movement. Smedley and Whiteley were among the thirty-one female chemists who got in touch with each other to compose a collective retort: they were, they wrote indignantly, united not politically, but by their love of chemistry. But then nothing happened until 1919, when the removal of the Sex Disqualification Act forced professional organizations to rethink their standard policy of discriminating on grounds of gender or marital status. Although many universities and other employers found ingenious ways round the problem, the Chemical Society did, at last, accept unanimously the proposal to admit female members.

Between them, around seventy men nominated twenty-one female candidates, who were elected as fellows of the Chemical Society in 1920. They included the ringleaders Ida Smedley and Martha Whiteley, who initiated a women-only dining society that met every four months at the London Lyceum Club established by Ida's sister Constance. Within a few years, Whiteley was on the Chemical Society Council, well placed to countermand the arguments of Armstrong and his allies. But Armstrong equivalents lived on at the Royal Society. In 1943, when Lawrence Bragg canvassed opinions about Kathleen Lonsdale, who two years later became one of the first female FRSs, one distinguished X-ray crystallographer replied: 'I must confess that I am one of those people that still maintain that there is a creative spark in the male that is absent from women, even though the latter do so often much marvellously conscientious and thorough work after the spark has been struck.'[33] And when another professor looked back from 1983, he remembered Ida Smedley MacLean not for her work, but for her association with two men: 'Mrs MacLean. Yes, she worked at Manchester. She worked for a time as Robinson's assistant and then she married MacLean.'[34]

PART IV

SCIENTIFIC WARFARE, WARTIME WELFARE

11

SOLDIERS OF SCIENCE

Scientific Women Fighting on the Home Front

> One to destroy, is murder by the law;
> And gibbets keep the lifted hand in awe;
> To murder thousands, takes a specious name,
> 'War's glorious art', and gives immortal fame.
>
> Edward Young, *The Love of Fame*, 1725–8

Aldous Huxley, author of *Brave New World*, once remarked that the *Encyclopaedia Britannica* defined three types of intelligence: human, animal, and military. Mabel Elliott corresponded to two of those categories. Although she was one of the thousands upon thousands of extremely intelligent women who had never benefited from a university education, she helped to transform military intelligence into a scientific study.

With so many men away fighting abroad, women seized the opportunity to demonstrate that they were capable of far more high-powered office work than typing. Fluent in four languages, Elliott languished as a high-powered secretary until the First World War offered her the openings for a meteoric rise in the Censorship Office. Suspicious of an apparently innocuous letter she was inspecting, she heated it gently to reveal secret messages inscribed in lemon juice. She realized that a German spy posing as an American wool merchant had been ostensibly contacting a business address in Holland, but in actuality had been passing on information about Britain's defence strategies. Her acuity led to other similar discoveries, so that eventually a complex espionage network was exposed and foiled.[1]

After that initial success, Elliott was repeatedly promoted until she ended up in charge of over 3,000 women. After the War, she opened her own secretarial bureau and began a new career at the Royal Society of Chemistry, where she became the first woman to be elected an honorary member. During the public trial of the spy ring, her name had been altered to conceal her identity, so presumably many people wondered why exactly she was made an MBE and why the French government awarded her the Palme d'un Officier de l'Académie. Elliott's spy-catching achievements only came to light in 2011, when archivists at the Royal Society of Chemistry carried out their own detective work and read between the lines of her personal records.

Similarly, official secrecy has obscured the wartime activities of the chemist Frances Micklethwait, who had collaborated in various research projects for thirteen years before she joined a research team at Imperial College investigating chemical weapons during the War. She was among the first to study mustard gas, a particularly pernicious chemical because, unlike chlorine, it is invisible and its effects are not immediately noticeable. A colleague later recalled noticing her and her fellow workers 'popping into the corridor every now and again to see how the experimental blisters on their arms were getting on'. Like Elliott and other scientific women, Micklethwait was decorated after the War, but because their work was top secret, its significance has only recently begun to be appreciated.[2]

Whereas men are celebrated as war heroes, women who carried out equally brave or significant tasks lie neglected. The ex-Girton mathematician Beatrice Mabel Cave-Browne was a schoolteacher for eleven years, but in 1916 she went to work for the government on aeroplane design. She produced two very important research papers, but they remained unpublished for fifty years, banned under the Official Secrets Act.[3] In Newnham's hand-lettered commemorative volume, an otherwise forgotten E. K. Lynch is reported as serving '5 months on the staff of the Cambridge Scientific Instrument Co, on work connected with a new secret invention for use in war'. The activities of other Newnham graduates are similarly mysterious. For instance, why was an 'unpaid clerk' awarded an MBE?[4]

Veils of secrecy still shroud geographical regions packed with women workers, such as Gretna and its sprawling munitions factory, or the area surrounding Bridport in Dorset, where the work of the village women was effectively suppressed, because—most unusually—the newspapers carried virtually no accounts of local wartime activities. Adapting their traditional skills of knotting West Country hemp and flax into ropes and fishing nets, these women spent much of the War making enormous quantities of military products that would nowadays be manufactured by machine—hammocks, food bags for horses, lanyards, safety rings for hand grenades. In nearby Taunton, their sisters and cousins wove over 8,000 miles of cloth for puttees to wind round soldiers' legs. A few months into the War, the Admiralty approached the Bridport companies for help with a still bigger requirement—they wanted strong wire nets that would be able to trap submarines as if they were giant rabbits or herring. Using hastily modified looms, the women solved this technological challenge by creating huge metallic curtains that could be draped across the Channel to catch enemy submarines.[5]

Forgotten for a century before their contributions were recently brought to light, even now these women remain mostly anonymous. But at least they can be collectively commemorated for their wartime defence work.

Deciphering Secrets

Cryptography was not in itself new, but this was the first war in which codebreaking played such a crucial role. The relatively recent techno-logical innovations of a global telegraph network and long-distance radio (then known as wireless) meant that essential military informa-tion could be communicated far more quickly than ever before, and it was vital for Britain to defend its own secret systems while sabotaging those of the enemy. Women played an increasingly important part in both aspects of this electrical warfare—developing and operating the technology needed for transmitting and intercepting messages, as well as the more analytical tasks of deciphering and assessing intelligence information.

Because Britain had already improved its communications systems before the War by investing in both cable telegraphy and wireless radio, the nation enjoyed a double technological advantage. Telegraphy had first been introduced during the first half of the nineteenth century, but by the time of the War, Britain had taken advantage of its imperial connections to dominate the underwater cable system that wrapped itself around the world. The twentieth-century development of radio meant that London could communicate directly and immediately with ships and soldiers instead of waiting while written instructions were sent from one place to the next. Of course, because anyone could tune in to the right frequency, every country rapidly developed its own ciphering system to protect its messages. But even before the War started, Marconi and British military organizations had been collaborating closely to produce new equipment and to set up signalling stations around the coast and overseas. As a result, radio operators could intercept German messages and relay them to London for decoding.

By the end of the War, Britain was relying on women to maintain the staff levels needed for operating the country's unparalleled modern electrical communications systems. Rather belatedly, in the last months of 1917, the Admiralty decided to establish the Women's Royal Naval Service (WRNS). Characteristically, their recruitment drive consisted of placing an advertisement in the *Times*. By then, so many high-powered women had taken up work elsewhere that the pool of available talent was small, and most of the new trainees were eligible only because they had just passed their eighteenth birthday. One exception was Vera Mathews, who had long been committed to naval work and had previously applied for a job with the Admiralty, only to be told 'We don't want any petticoats here.' Subsequently, officials had the grace to recognize their mistake: Mathews proved so successful after training that she rose right to the top of the WRNS during the Second World War. In 1918, the numbers of female naval recruits grew to almost 7,000, many of them working as electricians, codebreakers, and radio operators. Presumably the male officers who taught them wireless telegraphy were astounded when one women's class gained the highest marks ever achieved. And some of those women were equally astounded (as well as angry and disillusioned) when in 1919 they were dismissed with a week's pay and the Service was closed down.[6]

The day that War was declared, 5 August 1914, the British Navy cut all the German cables except the one that went via London. Although the Germans quickly retaliated by slicing through British links, they were forced to send diplomatic telegrams along commercial networks, parts of which were under British control. This forced the Germans to rely on radio—but unknown to them, their codes had been broken very early on by men and women working under conditions of high security in Admiralty and Army headquarters. To ensure that the Germans would not suspect their messages were being read, the strictest secrecy was maintained, a strategy that paid off. In 1917, British cryptographers intercepted what became known as the Zimmerman note, a proposal from Germany to Mexico that they should form a military alliance. Three months later, the United States of America declared war on Germany.

Some women were involved in this intelligence work, but their activities were kept so secret that even close colleagues knew nothing about it. Although rumours of an Admiralty intelligence unit in Room 40 did leak out, very few people were aware that another office, Room 45—officially known as ID25—even existed, let alone appreciated what its occupants were doing.[7] Initially it was staffed by naval officers, but the Navy began recruiting university graduates, both male and female. By the end of the War, a team of codebreaking experts—thirty-three women and seventy-four men—worked round the clock in shifts (technically known as watches in this Admiralty department), sleeping on the premises in makeshift beds. Intercepted German signals were sent up by pneumatic tube from the basement and distributed round the desks by wounded officers (who included the one-legged nephew of H. Rider Haggard). The Admiral in charge, Reginald Hall, was an unconventional man who succeeded by breaking rules. His top priorities were to maintain secrecy and identify innate wit, whether male or female. 'Men and women of every profession and class joined us,' he wrote in his gossipy memoirs that were judged too sensitive to publish, 'and in many cases they had little but their own good sense to guide them aright.'[8]

Like so many women in the War, intelligence specialists were—as Hall freely admitted—making it up as they went along. They had to acquire new intellectual and manual skills, but they also needed to invent ways of surviving in the nearly unprecedented situation of a mixed work environment. In this confusing new world where everyone was working for war, female university graduates—especially scientists—did at least hold the advantage of having attended mixed lectures or practical classes. Ideologically, if not always in practice, the laboratory was a secluded egalitarian space with its own conventions, where female identity was left behind at the door to the outside world. 'We have no airs and graces here,' said Wells's fictional biologist as she clicked shut her microscope case; 'my hat hangs from a peg in the passage.'[9]

Very few details survive, but the codebreaking women included Dorothy Gilliatt (who married one of her male colleagues, Alastair Denniston, in 1917), Miss Henderson (an Admiral's daughter), Miss Hayler (who specialized in Italian non-alphabetic ciphers), and Miss Robertson (headmistress of Christ's Hospital School for Girls and a Cambridge graduate). Other women were employed as secretaries, although the division of duties is not always clear. One of them was Olive Roddam, daughter of a distinguished military man, whose boss was Alfred Knox, in civilian life an expert on deciphering ancient Greek papyri. Roddam would arrive in the morning to find her manager soaking in a bath, the site of his most inspired intellectual feats—notably his success in picking apart a German flag code by deducing that a test message transcribed a famous German poem. After the War, Roddam and Knox got married, much to the disgust of Lytton Strachey, whose advances to Knox had been rejected.

Although valued for their skills and dedication, these women were also rated (as ever!) by their looks and behaviour. According to one of her male colleagues, Mrs Bayley 'caused slight but unimportant trouble by her very attractive appearance', as if the two men she had affairs with were innocent.[10] Among the secretaries, Lady Ebba Hambro, wife of a city banker, gained notoriety by smoking cigars, while Joan Harvey, daughter of the Secretary to the Bank of England, made herself popular by running the tea club. Reminiscing fifty years later, when identities could be revealed, one of the disabled male clerks said nothing about the women's expertise, but instead singled out Margy Bayley as 'one of the first ladies to "shingle" [a distinctive short hair style] her hair', apparently much to the consternation of her many male admirers. Similarly, high intellectual abilities scarcely feature in his accounts of Bea Speir, reportedly renowned for the variety of the bandeaux she wore round her head, and Rhoda Welsford, 'a beautiful blue-stocking and a magnificent dancer'.[11]

The male and female codebreakers of ID25 shared an intellectual quirkiness, an obsession with puzzles, twisted logic, and insider jokes. Confined within their cramped secret quarters, they developed their own internal codes to survive the intense pressure of knowing that their speed in unravelling a text could determine the safety of the nation. At the end of the War, Knox and a couple of his close friends were inspired by Lewis Carroll's *Alice in Wonderland* to write a witty parody of their life in this bizarre secret world where words never meant what they said. Performed as a private pantomime only once, *Alice in ID25* remained hidden and unpublished until 2015. Although never intended as literal truth, it does convey a flavour of the atmosphere in Room 40 and its neighbours.

In the very first line, Alice is walking down Whitehall after breakfast at the 'Ship', presumably where the cryptographers gathered after an all-night watch. Soon she tumbles down not a rabbit hole, but one of the pneumatic message tubes, landing up in a cage of golden wire before being dumped in a tin marked NSL—Neither Sent nor Logged. Alice is constantly perplexed by back-to-front writing, German words, acronyms, references to keys, and zany attempts to code her as Gertrude or a bun. The major characters—the White Rabbit, the Dormouse, the Mad Hatter—are all identifiable, and women also feature in her adventures. To make her grow, Alice is prescribed 'Harvey's Fatting Food' because Joan Harvey was in charge of providing tea, while Lady Hambro appears as 'a huge creature...conducting an orchestra of typewriters with the much-bitten end of a pencil'. Hambro was referred to as Big Ben, Alice is told, because she was so striking—and in an even more painful pun, because she 'strikes' again by mocking the Hatter who went to Harwich (a joke that must have made sense at the time). But although there are other standard female put-downs, such as 'little Miss Mouse', Knox (of the bath) also mocked himself as Dilly the Dodo, the 'queerest bird' Alice had ever seen: 'long and lean, and he had outgrown his clothes, and his face was like a pang of hunger'. In the final imagined scene of demobilization, there was presumably some undertone of reality in the characters' feigned terror at being released out into the normal world.[12]

In addition to the Admiralty codebreakers, there was a separate military unit based in the War Office, code-named MI1B and headed by Major Malcolm Hay, a severely wounded Scottish laird and former prisoner of war, who spoke several languages. Like his counterpart Hall at the Admiralty, Hay insisted on a free hand in recruiting, and the two men cooperated closely, in contrast with the frequent counterproductive antagonism between wartime departments. It was here that Oliver Strachey, Ray Costelloe's husband, at last found his niche. Just as at the Admiralty, codebreakers were hired through personal contacts rather than any rigorous selection process, and a friend suggested that Strachey fitted the job description perfectly: 'someone with an ingenious head for puzzles and acrostics to decipher code and piece together scraps of wireless messages.' After forty years of aimless drifting, he joined the War Office and went on to become one of their most senior cryptographers.[13]

The initial group comprised only five men, but in 1916 Hay was able to obtain funds for expanding his team after poor deciphering was blamed for disastrous losses during the Battle of Jutland. First there were ten women and ten men, but by the end of the War MI1B employed almost a hundred people and had moved into more spacious Mayfair premises. To ensure that new additions to his 'Cork Street Code Breakers' were of a sufficiently high calibre, Hay set a ferociously tough entrance examination based on a doctored French newspaper article. The only person who ever completed it was a woman, Claribel Spurling, who in civilian life co-authored books and plays for children. An old sepia wartime photograph shows her sitting between Mrs Scott and Mrs Edmonds, although it is unclear whether they were secretaries or fellow cryptographers.

The MI1B team shared ID25's idiosyncratic sense of humour. A few days after an unexpected and intrusive military inspection, one of Hay's codebreakers handed him an extract from an ancient document found in an Egyptian sarcophagus. Footnoted by Professor Sillirotter (try reading the names out loud), the codex describes Hamfês of Khâ-ki—the Slayer of the people of Bosh, who spoke naught to the Civ-il-yân tribe—and a scribe who testifies 'Yuk-an sertsh-mi'.[14]

Domestic Warriors

Long before the government realized how important women were going to be in this War, female volunteers had started organizing themselves into miniature battalions, each with its own distinctive uniform and military-sounding name—the Women's Defence Relief Corps, the Women's Auxiliary Force, the Home Service Corps. The NUWSS immediately set up employment bureaux and training workshops, while only a fortnight later, Queen Mary established a fund to help women thrown out of work; within six months, it had assisted 9,000 applicants in seventy special relief workrooms. Committees proliferated. Around the country, 2,500 were set up just to look after Belgian refugees, while others cared for diverse distressed ladies— musicians in search of concerts in which to perform, graduates who suddenly needed to support themselves as governesses or teachers, garden-owners with grander ambitions who wanted to tend farms.[15]

However well-intentioned these initiatives may have been, they were not necessarily well received. Cartoonists mercilessly lampooned the Women's Volunteer Reserve, founded by the Marchioness of Londonderry, who gave herself the rank of Honorary Colonel-in-Chief. Dressed in khaki, its members practised military drill in their bid 'to provide a trained and disciplined body of women' who could free up men by becoming despatch riders or telegraphists. These voluntary recruits were criticized for their mannish behaviour and their presumption in adopting a soldier-like uniform without being forced to confront the risk of death. 'Khaki is sacred to those for whom it forms a shroud,' pronounced *The Globe*; 'it is the garb of the fighting man...and it should not be lightly donned.'[16]

There were also mixed feelings about the privileged women who had previously been militant suffragettes but promptly switched to the other side of the law by converting themselves into police volunteers. Professing themselves shocked at the sight of women who seemed to be loitering around railway stations, they began monitoring the munitions factories as well as patrolling London's streets and parks in quest of any untoward behaviour. Unsurprisingly, they antagonized many people by their officiousness and their unrealistic standards of propriety, but the official line was different: these ladies were, it was reported, dramatically improving the welfare and morals of the lower classes. The sanctioned account of a philanthropic 'Superwoman' supervisor of 25,000 women at Woolwich Arsenal does stretch credibility: 'When Miss Barker enters, their faces light up, gay greetings pass, and one feels instinctively the confidence and mutual trust with which she has inspired her great family.'[17]

In contrast, after their initial reservations, government officials effusively welcomed women who undertook voluntary work—but perhaps unsurprisingly, they neglected to mention how much the country was benefiting by tapping into a large reservoir of free labour. Herbert Asquith, Prime Minister until he was replaced by David Lloyd George in 1916, revised his initial suspicions to praise women for going against their inherent feminine nature by performing such acts of patriotic self-service. Lloyd George delivered a similar message more eloquently: 'Without women victory will tarry and the victory which tarries means a victory whose footprints are footprints of blood.'[18]

After a couple of years, centralized public organizations began taking over much of this voluntary work. Initially, the government had awarded grants to operate training farms, but in 1917 the Ministry of Agriculture decided to coordinate individual local initiatives by launching the Women's Land Army. Similarly, the Ministry of Munitions assumed financial responsibility for the welding workshop that had been set up by London suffragists under the experienced metalworker Miss F. C. Woodward. As a result of this institutionalization, more structured opportunities to progress emerged for intelligent women who were of a scientific or technological inclination but lacked the means for a university education. Among the welding workshop's early students was Ethel Rolfe, whose first position was as a lone woman working in an aeroplane factory. After gaining this on-the-job experience, she found a new role as a government advisor, visiting factories around the country and recommending how women could be rapidly trained to take over production tasks. In 1917, this was described as a 'typical' career path for a woman who for the first time had been able to fulfil her girlhood ambition of entering an engineering trade. Among the Ethel Rolfe equivalents who have left little trace behind them was Margaret Godsall, a fellow welder so committed to her responsibilities that she died after staying late to get a rush order finished when she was suffering from flu.[19]

As the War continued, government departments expanded and new ones sprang up—the Ministry of Food, the Ministry of Information, the Air Ministry—established in temporary huts spread across London's parks and gardens. The censor's office had originally been housed in the General Post Office, but it spawned several subdepartments, run by 2,000 staff. By the time of the Armistice, the Ministry of Munitions had acquired seventy-four central London buildings and over 25,000 employees. Whereas Whitehall and the City of London had previously been dominated by dark-suited men, now the streets were thronged with young women wearing colourful dresses, an easy target for critics who slated their skirts as scandalously short, and their earnings as scandalously high.[20]

There was a mixed economy of female workers, some of them fully employed but many either volunteering with no pay or privileges, or provided with only temporary contracts. Even those who served abroad with the Army were never fully enlisted like their male colleagues—they were enrolled instead, often through labour exchanges, and so had the status of civilians in uniform rather than being fully fledged military personnel. Almost from the beginning, nurses (but not female doctors) were hired to work both at home and abroad, and they were supplemented by untrained women who signed up for the Voluntary Aid Detachments (VADs). In 1916, the government began enrolling cooks, motor drivers, and mechanics, and the following year, women-only army and navy units were established so that soldiers could be released from routine tasks both at home and abroad. In addition, many thousands of non-enrolled women worked on a semi-permanent freelance basis—as exploitative as modern zero-hours contracts, and similar to the situation in which Dorothea Bate and her museum colleagues found themselves. These women carried out a variety of vital technical tasks: making and repairing gas masks, testing radio valves, photography. Others with good mathematical skills were set to organizing payrolls or introduced to forestry, which entailed measuring trees to calculate how much wood they could yield, as well as sawing them down.

Some university scientists took on senior administrative positions in government organizations. Ethel Thomas, for instance, was head of the botany department at Bedford College, but she also started working for the Medical Research Committee (not yet the Council) and became a Woman's Land Army Inspector. The College botanical garden she had designed was opened just before the War, in 1913, at their site in Regent's Park, but she was evidently a difficult woman to work with. Three years later, she had antagonized so many of her colleagues that she was fired, but she went on to follow a very distinguished academic career and was a keen supporter of the BFUW.[21]

In contrast, the Newnham geologist Gertrude Elles seems to have been far more amenable. Renowned for her brown tweeds and russet felt hat that swept down across her nose, Elles acted during the War as the Red Cross Commandant at the local hospital and was rewarded with an MBE. An exceptionally talented research scientist, she had been awarded research funding from the Geological Society—but imagine her frustration at not being able to receive it formally because, as a woman, she was banned from attending meetings.[22] Reminiscing on her time as an undergraduate, a former student remarked that 'The energy, kindliness and enthusiasm of Miss Elles made her perhaps the most inspiring influence in those glorious years.'[23] Elles collaborated with another Newnham graduate, Ethel Shakespear, on a monumental monograph about graptolites, a type of fossil. A star pianist and tennis player as a student, Shakespear had gone to Birmingham University in 1896, but both during and after the War she worked for the Ministry of Pensions, campaigning to secure care and retraining for disabled soldiers.[24]

Once you start looking for them, such iceberg-tip mentions appear surprisingly frequently, although disappointingly few details survive to fill out lives that were so rich in energy and initiative. For instance, a woman called Gertrude Shaw had been the headmistress of a girls' high school in Leeds, but in 1913 she started working at the Women's Institute in Woolwich. Soon she was running a canteen for the Arsenal's munitionettes, and then she was put in charge of a large residential unit for 6,000 factory workers at Coventry. By the end of the War, she was travelling all over the country as a Ministry of

Munitions Inspector. Was she the same Gertrude Shaw as the woman who released suffragette leaflets from the top of the London Monument and was secretly photographed in Holloway prison while she was on hunger strike?[25] Perhaps. Or perhaps even probably. But will we ever know for certain?

There are other shadowy examples of wartime scientists, such as the physiologist Marion Greenwood, who was active throughout her life as a suffragist and educational reformer, and served on the Women's War Agricultural Committee. She had previously run Newnham's Balfour Laboratory for nine years and seemed all set for an eminent scientific career until in 1899 she got married, and abandoned teaching and research—but although ostensibly dedicated to her tubercular husband and her children, she still carried out a great deal of unpaid voluntary work.[26] Margaret Fishenden also got off to a brilliant start, but her career went in a very different direction. After studying physics at Manchester, she headed their meteorology department for five years, specializing in air pollution. During the War, she resigned from this academic position and began directing the city's efforts to clean up its smoky atmosphere. Although she later went to Imperial College, it was this wartime industrial experience that enabled her to become a renowned expert on domestic coal fires, at that time a nationally important topic. Yet despite her scholarly prowess, and despite the economic value of her research, Fishenden was only an honorary lecturer at Imperial. After her official retirement she went on lecturing, presumably because as a single mother, she had not been able to build up an adequate pension.[27]

By 1914, thousands of women had studied at British universities in what are now known collectively as the STEMM subjects—science, technology, engineering, mathematics, and medicine. What happened to them all? What did they do during the War? For most of them we shall never be able to answer those questions. But the surviving evidence of a few does suggest that many graduates played active and important roles during the War and beyond.

12

SCIENTISTS IN KHAKI

Mona Geddes and Helen Gwynne-Vaughan

'...One of my mates is dead. The second day
In France they killed him. It was back in March,
The very night of the blizzard, too. Now if
He had stayed here we should have moved the tree.'
'And I would not have sat here. Everything
Would have been different. For it would have been
Another world.'

Edward Thomas, 'As the Team's Head-Brass', 1916

According to the *Scots Pictorial*, when Mona Geddes became the first woman to graduate in medicine at Edinburgh, 'The men took the innovation gallantly...she was greeted with the very heartiest applause.'[1] Whether their standing ovation was sincere or not, almost twenty years later she achieved another first—but this time men responded with far less gallantry. After she was put in charge of organizing the brand-new Women's Auxiliary Army Corps (WAAC), many traditionalists reacted rudely, openly expressing their contempt of women who entered the male sphere of military matters.

While Geddes supervised affairs in Britain, a university botanist called Helen Gwynne-Vaughan established the women's units in France. Already familiar with discrimination in the laboratory, she knew how to cope with disdainful behaviour. When she politely asked a colonel if his wife was interested in this new military initiative, he retorted tetchily: 'My wife, Mrs Gwynne-Vaughan, is a truly feminine woman.' Herself a glamorous dresser, Gwynne-Vaughan winced, but diplomatically remained silent and persevered in her new job. Military men, she later concluded, regarded WAACs in one of two ways—either as a uniformed soldier who happened to be wearing a skirt, or as an incompetent in petticoats. Presumably the tactless colonel enjoyed laughing with fellow officers at the standard joke that she kept on hearing and that profoundly irritated her: 'Would you rather have a slap in the eye or a waac on the knee?'[2]

At the beginning of the First World War, it was pretty well unthinkable that women should be involved in military work, especially overseas and near a zone of action. But as casualties mounted, and as women were proving their capabilities in factories and offices all over Britain, the government decided to launch a recruitment campaign for female troops. Clearly, more soldiers would be available to fight at the front if women took over camp-based roles such as cooking, cleaning, and driving, but sceptics questioned their commitment, ability, and endurance. Whereas the government slogan declared that 'Every fit woman can release a fit man', the military commander Douglas Haig insisted that 200 women would be needed to replace not 200, but 134 men. Objectors also worried about women's damaging impact on both morals and morale. Their responsibilities were carefully delineated: even if women were to be allowed out of their own homes and behind the lines, they were to carry out domestic-style tasks and be prevented from close contact with soldiers. Haig's stiff pronouncement that 'Clothing storekeepers cannot be women for they cannot assist at the trying on of clothes' is laced not only with etiquette-laden horror but also with suspicion.[3]

After much negotiation, the WAAC was founded in 1917. The Minister of War, Lord Derby, favoured appointing titled ladies to run the new service, but he was overruled by his staff: they preferred to co-opt a professional woman for this demanding position, and they had the perfect candidate—Mona Geddes, who just happened to be the sister of the Director of Recruiting at the War Office. Inevitably, there were heated debates about the women's behaviour and appearance, but after many negotiations, WAAC recruits in khaki started to become a familiar sight on British streets. Daughters trapped at home under strict parental control eyed them enviously. 'Then my evil spirit and restlessness began to catch up with me,' Lia Parfitt reminisced. 'Unbeknown to my father I wrote to the Recruiting Office of the WAACs and offered my services. Soon a long envelope bearing the magic letters OHMS came for me.... My father would happen to be at home at the time and he hit the roof!'[4]

Both given the title Chief Woman Controller, Geddes and Gwynne-Vaughan confronted countless unanticipated obstacles, many of which they regarded as laughably petty and/or insulting. For example, as the nation's employees, the female forces expected an official military uniform that would differentiate them from volunteers, but the Army cavilled at the expense of providing khaki that was already in short supply and seen as more suitable for men. In contrast, sceptics insisted that the WAACs should be readily distinguishable from prostitutes trying to enter the camps. Although the quasi-military voluntary organizations had been criticized for undertaking the masculine performance of drill practice, the WAAC leaders held firm against attempts to ban it, pleading its necessity for physical exercise and hygiene.

And then there was the badge problem. In principle, soldiers should salute higher-ranking officers, but members of the WAAC were technically civilians, so that was deemed inappropriate. But the female leaders also had a strong argument: in order to maintain discipline, surely it was essential that the hierarchical military structure be made apparent and be observed? This impasse was resolved by a compromise. Whereas the ranks of male commanders were denoted by crowns and bars, status in the WAAC would be signified by suitably feminine flower symbols. Eventually, Geddes and Gwynne-Vaughan were allowed to wear a double rose, informally equivalent to the crossed sword and baton of a brigadier-general.[5]

Both of them scientifically trained—one a doctor who published books and papers on nutrition, the other a botanist specializing in the genetics of fungi—Geddes and Gwynne-Vaughan were temporarily diverted from their own careers to run a female military organization that only three years earlier would have been impossible to imagine. By the end of the War, between 80,000 and 90,000 women had enrolled in units attached to the Army, the Navy, and the Air Force, the great majority in the WAAC. Their collective label—the auxiliary forces—implicitly underlined their inferior and temporary position.

Sent back into civilian life once the War had ended, these women were expected to resume their place as second-class citizens. From 1918, any man who had served in the military automatically had the right to vote, but the female franchise was restricted, most significantly by age. The nation seemed eager to forget that these women had confronted danger, had worked for long hours at low pay, and had endured appalling living conditions. Article after article criticized former WAACs for continuing to smoke and thrust their hands into their pockets. Regretting 'the loss of grace and charm which in the old days caused their fathers to espouse their mothers', antagonistic journalists expressed their 'hope that it will represent but a passing phase and that [they] will venture to shed their army manners with their military uniforms'.[6]

Alexandra Mary Chalmers-Watson
(aka Mona Geddes) (1872–1936)

To start with, a technical aside...

Confusingly, this woman is referred to by several names. She was christened Alexandra Mary Geddes, but seems to have abandoned Alexandra except for official occasions; most people called her Mona, an affectionate diminutive introduced by the ayah who came back with the family from India to Scotland when her charge was four. The very afternoon of her graduation, she married another doctor, and was transformed from Miss Geddes into Mrs Douglas Chalmers-Watson, although, as a further complication, some of her contemporaries dropped the first part to call her Mrs Watson.

And now for some more interesting aspects of her life…

In retrospect, Mona Geddes's fate seems to have been predestined from birth. Her mother's sisters-in-law included two of Britain's most eminent campaigners for female equality. One of them was Elizabeth Garrett Anderson, the first woman to qualify as a medical doctor in Britain and co-founder of the first British hospital for women; a surviving news film shows her as an elderly woman leaning on a stick as she marches at the head of the massive procession of 1910.[7] The other was Millicent Fawcett, the leading suffragist who became President of the NUWSS and co-founded Newnham College. Among Geddes's own siblings were a general who later became First Lord of the Admiralty, and a former anatomy professor who was appointed to the War Office. And on top of all that, her mother Christina had helped to establish the Medical College for Women in Edinburgh. Hardly surprising, then, that Geddes was not only a doctor but also the first head of the WAAC. Indeed, when she was twenty-six, an eccentric aunt reported summoning up her psychic powers to single Mona out as the diamond among her siblings:

> When she comes into the room I feel as if the lights had gone up. We are all proud of her for the way in which she has blazed the trail of medical education for women in Edinburgh. Her achievement is magnificent but she will rise to greater heights in years to come.[8]

Growing up in Edinburgh, Geddes was introduced to science by her father, who taught her basic astronomy, agriculture, and engineering. When she was fourteen, the family visited the Edinburgh Exhibition, where she was particularly fascinated by displays of food preparation and preservation. Nutrition later became her specialist research topic, an interest fanned by her mother's regrets that privileged women with servants had become divorced from traditional culinary skills. The following year, 1887, her mother became involved in establishing the residential Edinburgh School of Cookery and Domestic Economy. With its twin aims of improving diet and educating women, the

School expanded over the decades to become the present-day Queen Margaret University.

For her next project, Christina Geddes turned to medicine. Closely related to two women doctors, and despite being chronically ill, she helped to set up the Edinburgh College of Medicine for Women, which opened in 1889. Back in the early 1870s, a few women had tried to become doctors in Edinburgh, but had found it impossible. In particular, Sophia Jex-Blake had gone south to found the London School of Medicine for Women before qualifying in Switzerland and Ireland. Returning to Edinburgh, she opened her own medical school in 1886, but it soon ran into financial and administrative difficulties. Following an unpleasant court case, several disgruntled students went elsewhere, including Elsie Inglis, who later became Scotland's most celebrated female doctor of the First World War. Displaying the initiative and obstinacy that characterized her entire career, Inglis launched an alternative training centre for Edinburgh women. Her powerful allies included Christina Geddes, who enlisted the help of two relatives, both doctors—her sister Mary and her sister-in-law Elizabeth Garrett Anderson, who had known Jex-Blake for about thirty years.[9]

Their rival college opened in 1889, and it flourished, effectively forcing Jex-Blake's earlier school to close down. Although a few male doctors were extremely supportive, a solid core of diehard resisters made the women's lives very difficult. Disapproval was focused particularly on the propriety of men and women attending physiology classes together, and it was only in 1916 that the War persuaded Edinburgh into allowing female students to attend all the lectures and clinics. Mona Geddes must have endured experiences similar to those of a student at a London medical school that had reluctantly gone mixed: at the inoculation department 'she had to deliver a specimen there for examination and was warned to knock but not enter and wait until someone came to hand over to. The consequence was that we learnt very little bacteriology and no immunology.'[10]

In order to become a fully qualified doctor, Mona Geddes had first to graduate in medicine, and then to write a dissertation, for which she spent about eighteen months working as a medical officer in the East End of London, mostly at a maternity hospital but also at an orphanage. On entering the city slums, she encountered lives far removed from her own sheltered existence in Edinburgh—children who were sick and abandoned, babies dying from malnutrition, houses swarming with vermin. Finally, in 1889, she walked across a platform as the first woman in Edinburgh to receive the coveted cap of a doctor. Only a few hours later, she had discarded her academic gown to don her wedding outfit.

For the next sixteen years, the Chalmers-Watson couple led a busy professional life. In addition, by the start of the War, they already had two small sons, aged seven and three. Douglas Chalmers-Watson adopted the convenient role of absent-minded professor, which meant that his wife (like many others) ended up organizing the household as well as much of their collaborative research into the effects of trace substances in food (later named vitamins). As with so many husband–wife teams, who did what is unclear, but together they edited the fifteen-volume *Encyclopedia Medica*: published under his name only, it is now seen as such a seminal work that in 2012 it was reissued as a paperback series; she also worked with him on two further specialized books about nutrition. After the War, the couple inherited a farm and set up a model dairy. As a student, Mona Geddes had been fascinated by the latest research into tuberculosis, a disease then responsible for many deaths. Although the bacillus had been isolated in the 1880s, uncertainty remained about how it was transmitted and the extent to which dirty milk was responsible. The couple became famous across Europe for their pioneering experiments on breeding cattle and producing safer milk.

Intending to be complimentary, an anonymous obituarist summed up Mona Chalmers-Watson as an all-rounder who would have been a great physician if she had not undertaken so much else.[11] Inspired by Millicent Fawcett and motivated by her spell in London, she was determined to improve the position of women and the life of the poor, and was active on an extraordinary number of committees—the Patriotic Food League, the Women's Emergency Corps, the Scottish Women's Medical Association, the Queen's Institute for District Nursing. She was involved in the foundation of the Queen Mary Nursing Home and the Elsie Inglis Hospital for Women, and herself worked as a physician in a hospital for women and children. On top of all that, she was an active suffragist, and became the first President not only of the Scottish Women's Hockey Association but also of the Edinburgh Women Citizens' Association.

Several testimonies survive emphasizing Chalmers-Watson's empathy, inner serenity, and commitment to helping others. Attempting to describe her, a journalist opened by praising her conventionally feminine characteristics: 'No whirlwind, no electric force, she is a quiet, smiling, tactful, and clear-headed woman of the world…'. The second part of his accolade explains her professional success: '…getting her way because her way is clearly the right way, never fussing people, never losing her head'.[12] Shortly before she died from radiation poisoning acquired during her years of research, Chalmers-Watson declared that her whole life had been haunted by the concept of duty, which had been dinned into her from an early age. When the War started, she felt torn by three conflicting duties: to her family, her scientific research, and her country. Despite all her medical and charitable work, she chastised herself for not having done enough since leaving London to help those working men who were volunteering for their country but had received so little from it.

Determined yet diplomatic, and prompted by her sense of duty, she drew up plans for a woman's military force, and, after some deft manoeuvres by her influential brother, was persuaded into accepting the task of heading the WAAC. From the beginning of 1917, Chalmers-Watson somehow managed to appear a conventionally committed wife and mother while at the same time setting up from scratch a new organization that operated with military efficiency. Although unhappy about leaving her family, she moved down to London for a year but drew the line at travelling to France, where the greatest number of women would be going. She turned for advice to her school friend and cousin Louisa, daughter of Elizabeth Garrett Anderson. An ardent militant suffragette who had risked her professional reputation as a doctor by throwing a brick through a window, Louisa had also been among the first to qualify at the London School of Medicine set up by Jex-Blake. She immediately suggested the perfect candidate for heading the French contingent—a university botanist, Helen Gwynne-Vaughan.

Helen Gwynne-Vaughan (née Helen Fraser) (1879–1967)

In 1917, Helen Gwynne-Vaughan was in a very different emotional place from her fellow Scot Mona Chalmers-Watson. She was thirty-eight years old, her husband had died of TB eighteen months earlier, and she had no children. Keen both to escape from her loneliness and to serve her country, she had already undertaken a course in medical bacteriology and had contemplated taking a mobile laboratory to a dangerous war zone. One Sunday evening in February, she was unlocking the door to her London flat when the telephone rang. In response, she rushed round to the Endell Street military hospital for women, where Louisa Garrett Anderson was the senior surgeon, and there she met Chalmers-Watson to learn that a woman's corps was going to France. 'It seemed like the realization of a dream,' she reminisced; 'I asked at once to be allowed to serve.' Only later did she realize that she would head the unit.[13]

Whereas Mona Geddes's trajectory through life might seem to have been mapped out in advance, Helen Fraser's was structured by dramatic turns of fate. Born a suggestive nine months after her father was called up to fight a war against Russia that never transpired, she was only five when he died from typhoid after drinking infected water in an Italian hotel. Until she was seven, the family believed that she would inherit an aristocratic title and estate, but after her uncle's sudden marriage, she was demoted far down the line of succession. And when her stepfather joined the Consular Service, she found herself moving from country to country—Corsica, Sweden, Portugal—and spending summers in Ayrshire with elderly relatives.[14]

Although nobody consulted her in advance, when she was fifteen Fraser was sent to board at Cheltenham Ladies' College. Just as abruptly, she was removed a year later, and thrown into the social whirl of balls, visits, and tea-parties deemed normal for aristocratic young ladies in search of a husband. It was not until 1899, when she was almost twenty-one, that Fraser won the family battle she had been waging for years—to study at university. The rows around the dinner table had been intensified by her insistence on studying the unladylike subject of zoology; still worse, she wanted to go to London, not to Oxford or Cambridge. When her family went back abroad in the autumn (this time to Algiers), she was allowed to stay behind and enrol in King's College.

As she must have anticipated, many further struggles lay ahead. To start with, the ladies' department did not teach zoology, but she eventually gained permission to follow the men's medical course and to take botany with animal physiology as a subsidiary. As another disillusionment, when she was awarded a special medal for coming top in botany, there were protracted discussions about whether a woman could receive it (again, she won). Day-to-day life was also difficult. Intimidated by the strange environment, Fraser was frightened to ask the meaning of scientific terms such as the buccal cavity (mouth). As there were no campus washrooms for women, on her way home she inspected her reflection in shop windows to make sure her face was free from splattered blood.

But in compensation, she revelled in the opportunity to study and to spend her days in a biological laboratory that must have been similar to one described in a fictional but nearby London college. It

> had an atmosphere that was all its own...long and narrow, a well-lit, well-ventilated quiet gallery of small tables and sinks, pervaded by a thin

smell of methylated spirit and of a mitigated and sterilized organic decay. Along the inner side was a wonderfully arranged series of displayed specimens.... [After their lectures] the students went into the long laboratory and followed out these facts on almost living tissue with microscope and scalpel, probe and microtome, and the utmost of their skill and care....[15]

Although she resented her mother's interventions, it proved invaluable to be backed by a prestigious family with useful connections. Fraser was introduced to the Keeper (department head) of Botany at the Natural History Museum, who found her a position as a part-time assistant and introduced her to influential scientists at the Royal Society. Taking advantage of such contacts, in July 1905, when she was twenty-six years old, Fraser secured a position at Royal Holloway College as a demonstrator, and she was later promoted to assistant lecturer. Elsewhere, she would automatically have been branded an outsider because she was a woman, but in this close-knit, all-female academic community she antagonized people by refusing to conform. Instead of the dark, unobtrusive clothes adopted by other women, Fraser favoured elegant, impractical costumes, even risking her job interview by wearing a brown linen dress and a hat covered with nasturtiums. Reportedly self-assured to the extent of arrogance, she deterred colleagues of her own age but charmed older male professors into acting as helpful patrons—'a stunner', 'a peach of a girl', they called her.

As a well-dressed scientist, Fraser challenged stereotypes. Even today, professional women can feel that they are expected to dress sensibly—that is, look dowdy—if they are to be taken seriously. How much more difficult it must have been for Fraser to negotiate boundaries and expectations. Brought up as a conventional socialite, Fraser had been taught to manipulate men through her looks rather than her intellect. She tended to inspire awed admiration from her colleagues rather than develop close working relationships. Academically, she began to wing it, to cut corners, an accusation levelled at her throughout her scientific career. Confronted with awkward questions from her students, she enlisted one of the college gardeners, her friend Evelyn Welsford, to look up the answers on her behalf. Insufficiently prepared for her doctorate, she failed at the first attempt, having relied too much on the support of a friendly fungi expert at the Natural History Museum, V. H. Blackman, who was also one of her examiners. Although Fraser had co-authored several papers with him, she was riled by his frank assessment of her academic ability.[16]

Her pride stung, Fraser threw herself into a new piece of research, obtained her DSc, and by the age of twenty-eight had become a lecturer in botany at the University College of Nottingham, living in her own flat and cared for by a servant. One of her students was D. H. Lawrence, who gained a distinction in Botany; neither of them reports paying much attention to the other, so it may or may not be significant that his character Ursula Brangwen is a schoolteacher and a botanist.

Fraser gained a reputation for being aloof, aggressive, oversensitive to criticism, and determined to further her own career. Nowadays, most men would treat her on her own terms as an equal, but her contemporaries were as perplexed as she was about how to behave. When she spoke bluntly, they answered politely—and so she often got what she wanted. After she learnt at a conference that the fossil fern expert David Gwynne-Vaughan was applying for a chair at Belfast, she had no qualms about canvassing support and negotiating herself into his botany lectureship at Birkbeck College (Figure 12.1). With the help of his recommendation, she took over his job; two years later she achieved her other goal of acquiring him as a husband.

Unlike many women, Fraser had no intention of ending her professional career once she had become Mrs Gwynne-Vaughan. Both partners were committed to their work, and they shrugged off family criticisms to live separately for a couple of years and fulfil their professional obligations. Perhaps the secret assignations, concealed from their colleagues, gave their relationship an added thrill, or perhaps they both needed time to adjust. Judging from contemporary comments, she was ecstatically happy, commuting between Belfast

Figure 12.1. Portrait of Helen Gwynne-Vaughan, by Philip Alexis de László, 1910.

and London, where she thrived both as an academic scientist and an active suffragist. With Louisa Garrett Anderson, she started a University of London Suffrage Society, although she refused to engage in militant activity and was more interested in women's economic independence than in their right to vote.

Inevitably, after a couple of years this charmed existence no longer seemed idyllic. Now she wanted to spend more time with her husband, and she wanted to have children. Maternity leave was unheard of, and the few women who did manage to combine motherhood with a scientific life—Marie Curie, Hertha Ayrton, Ida Smedley MacLean— were researchers rather than salaried lecturers. As a palaeobotanist, David Gwynne-Vaughan was a friend of Marie Stopes, and although it is unclear whether the two women knew each other, they evidently shared similar ideals, since Helen Gwynne-Vaughan tried (but failed) to set up an insurance scheme for covering salary losses during a professional woman's pregnancy. She also failed to become pregnant, and for once, failed to get a job that she wanted: although she applied to demote herself to a demonstrator so that she could work in her husband's Belfast laboratory, by then she had irrevocably alienated his colleagues through her refusal to behave like a conventional faculty wife.

Worse was to come: in September 1915, David Gwynne-Vaughan died of consumption. Like many mourners who are left behind, his widow suppressed her grief and submerged herself in time-consuming tasks, such as moving house and completing her husband's articles. This woman who had always been a lone outsider now found herself thrust into the ever-expanding community of wives whose husbands had died on service, easily identifiable in London's streets from their black dresses and veils. There was an immense surge in spiritualist beliefs during and just after the War, when wives and mothers tried to establish contact with husbands and sons whom they would never again see alive. Spiritualism is now ridiculed, but in the early twentieth century belief in another realm was endorsed by several eminent scientists, including Pierre Curie, the evolutionist Alfred Russel Wallace, and the physicist Oliver Lodge. Scientifically, it seemed feasible that an evolutionary process might lead to entities somehow accommodated within an invisible electromagnetic ether pervading empty space. Helen Gwynne-Vaughan was far from alone in seeking paranormal comfort. Ever the meticulous scientist, she carefully recorded her experiments with automatic writing and her experiences of posthumous visits from her husband.

For the next eighteen months, Gwynne-Vaughan stumbled on, burying herself in activity and determined not to let down her students. After she had taken her wartime medical course in bacteriology, she could find no suitable post abroad—and because she regarded herself as a professional scientist, she was too proud to behave like a typical middle-class woman and volunteer. In February 1917, the unexpected telephone call from Mona Chalmers-Watson offered her the ideal opportunity both to escape from her personal dilemmas and to act patriotically.

Many years later, Gwynne-Vaughan identified the characteristics that made her the right woman to take on this unprecedented

Figure 12.2. Portrait of Helen Gwynne-Vaughan, by William Orpen, 1918.

responsibility of establishing the WAAC in France (Figure 12.2). The first item she listed in her self-description reveals much about military organizations of the time: 'She was a soldier's daughter, which offered hopes of the right sort of heredity, she came from the same kind of family as the officers of the pre-1914 Regular Army and presumably spoke the same language.'[17] Within a month, the appointment had been confirmed and she was off to France, but beforehand there were important details to be worked out for this uncharted territory. Titles and rank were important. Conferring with her immediate superior Chalmers-Watson, Gwynne-Vaughan opted to be called 'Ma'am'; deviously, she managed to remove the 'Woman' from 'Chief Woman Controller' by pointing out that she did not wish to be known as the 'Chief WC.' After a visit to Marshall & Snelgrove department store, the two women designed a khaki uniform with no breast pockets (a mistake in practical terms, if not aesthetically), a daringly short skirt twelve inches from the ground, and a cap with a little veil at the back.

She later recalled how she felt on leaving England in March 1917. The day was misty and very still. Crossing the Channel one seemed to pass through dissolution into a new and different world. How different I did not at first realise.'[18] She did, however, notice one similarity: armies and universities were both hierarchical institutions, so she could simply transfer to officers the deference she had previously shown to professors. Within a couple of weeks, the first contingents of workers began arriving, and she spent much of her time trying to convince recalcitrant officers not only that her women were capable of hard work, but also that they should be allowed to salute, were not prostitutes, and were so determined to be treated on equal terms that they would donate their comfortable beds to the nearest hospital. Even so, she liked to observe the proprieties, returning from a brief trip home laden with silver cutlery and silk cloths, and accepting a ban on after-lunch cigarettes because 'Women in uniform, above all others, should avoid the suggestion of mannish behaviour.'[19]

Gwynne-Vaughan's published memoirs are packed with factual administrative details of her constant struggles to attain parity for her women. Occasionally, she allows her personality to flash out. She paid great attention to pay scales for different ranks, meticulously renegotiating the amounts that should be deducted to cover supposed extras such as food, laundry, uniform, and accommodation. Unsurprisingly, right through the War the civilian WAACs felt that they were worse off than the Army soldiers—but the soldiers were sure the unfairness lay the other way round. By obstinately sticking to her principles and instilling strict military codes of conduct, she made her own life even more difficult than it need have been, but her distress at being ordered to leave after a couple of years and take over the struggling Women's Royal Air Force (WRAF) confirms how deeply she committed herself to the work as well as to the women she commanded. Conversely, her troops showed great loyalty. One woman remembered meeting her: 'I was ushered into the presence of a tall, beautiful woman who met me on the easiest terms although my rank was of the lowest.'[20]

Gwynne-Vaughan understood at a personal, emotional level that 'Women whose husbands had been killed found comfort...in serving with the Army in which they had served.' One episode gives a particularly poignant impression of her loneliness, compassion, and effectiveness. Perhaps remembering her own pre-war desire for risk, when a recent widow 'begged to be sent to a dangerous post... I despatched her to an Army School with the assurance that it was frequently bombed.' The woman later became engaged to a soldier and had to be moved, but Gwynne-Vaughan commented without voicing any self-pity that she could not 'altogether regret that she had taken a second lease of life'.[21]

By the end of the War, Gwynne-Vaughan had become Dame Helen, the first woman to be made a Military Commander of the Order of the British Empire. An increasingly eminent public figure, she worked for several women's organizations and three times stood enthusiastically but unsuccessfully for Parliament. The women's military auxiliary forces were more or less closed down between the two Wars, but she remained a formidable fighter for equality in battles waged around committee tables. Although forced to accept that ex-service women would be paid less unemployment benefit than men, Gwynne-Vaughan adroitly managed to reject suggestions that their dependants should also have lower allowances: after all, she wittily pointed out, the dependant might be a boy. As committees proliferated in the interwar period, she was increasingly recruited by government organizations to advise on women police, food prices, and female unemployment.

Her academic career went less smoothly, partly because she gained a reputation for being awkward and arrogant. Reluctant to stir up memories of her marriage by going back to Birkbeck, Gwynne-Vaughan applied for a professorship at Aberdeen, but a man was appointed instead. He may well have been the better candidate, but her perception was of discrimination. After resigning from the WRAF at the end of 1919, she could see no alternative but to resume her pre-war job as Head of Birkbeck's Botany department, and two years later she was promoted to become the College's first female professor.

Open in the evenings, Birkbeck College catered for people with day jobs, many of them ex-servicemen regarded as the nation's saviours.

Gwynne-Vaughan's department was crammed into a narrow red-brick building in Chancery Lane surrounded by printing presses, and could be reached only by tiptoeing through other lecture rooms. Even so, she managed to draw in a series of distinguished botanists who worked there part-time, thus giving her students access to experts in their field as well as first-class experimental facilities. Abandoning her former fine clothes, for teaching she wore a white laboratory coat over a shirt and tie, and tried to blend in at committee meetings by donning dark suits and plain blouses. Determined to improve the College's low standing, Gwynne-Vaughan deployed the organizational skills that had made her so successful as a military leader, but which jarred in an academic setting and contributed to her swelling unpopularity. An exacting teacher who demanded the same high standards from her students that she imposed upon herself, she put enormous effort into supporting those who were ambitious and hard-working, but had little patience with others she judged to be shirkers. Any hint of insubordination was firmly quashed. Sitting in her small office, she monitored her assistants by listening to them lecture through a small hole cut in the dividing wall, and she became almost obsessively concerned to maintain order and cleanliness.

At first, she received several academic accolades, including an honorary doctorate from Glasgow, but she was gradually eased out of university committees. In response to her articles, Gwynne-Vaughan was often criticized for leaning too heavily on her students' research, and for clinging to out-of-date beliefs. Internationally, she was renowned for her 1927 university textbook on fungi, which was much admired for its clarity and comprehensiveness. But it was also slated for being too didactic, for informing its readers rather than encouraging them to think, a top-down approach that corresponded to her strict departmental regime. Her splendid career slowly fizzled out. Lonely and feeling herself to be the victim of repeated rejections, she increasingly turned in on herself to the point of reclusiveness or even eccentricity, eating on her own and isolated by deafness. By the time of the Second World War, she was no longer active in research, and after an inauspicious start, she was summarily replaced as commandant of the Auxiliary Territorial Service instruction school.

Distinguished as a scientist, a military officer, and a public figure, Gwynne-Vaughan was an extraordinary woman who remained active until a few years before her death in 1967. Was she right to feel that she had been unfairly excluded from even greater success? Inevitably, the few surviving accounts were written subjectively and according to expectations of the time, so it is impossible to know whether she was unreasonably uncooperative, or whether she was just being efficient. She may well have transgressed accepted norms of politeness for women, but on the other hand she was employed in a role that demanded masculine capabilities. Whereas a man might be admired for being authoritative, a woman behaving in the same way could (and still can) be judged authoritarian. She oscillated between tender compassion and offensive severity, displaying immense generosity towards individual students but reprimanding others abruptly for minor misdemeanours. She disliked flowers in the house because she was worried that the plants might have suffered from being cut, but thought nothing of upsetting anyone who gave them to her by snapping: 'Might as well have a lot of dying insects pinned up about the place as decoration.'[22] She was committed to improving working conditions for women, but retained old-fashioned convictions of upper-class superiority.

Was Gwynne-Vaughan a curmudgeonly disciplinarian, or was she simply determined to show that a woman can run an organization as effectively as a man? Perhaps more than any other woman of her generation, as a scientist and a military officer she had ventured into two male-dominated territories and in each one had fought her way to the top, where women as well as men resented her success. The life of a pioneer is inevitably hard, and she suffered.

13

MEDICAL RECRUITS

Scientists Care for the Nation

And for five weeks, ever since I knew that I must certainly go out with this expedition, I had been living in black funk; in shameful and appalling terror. Every night before I went to bed I saw an interminable spectacle of horror ... [but now] ... It is all unspeakably beautiful and it comes to me with the natural, inevitable shock and ecstasy of beauty. I am going straight into the horror of war. For all I know it may be anywhere, here, behind this sentry; or there, beyond that line of willows. I don't know. I don't care. I cannot realize it. All that I can see or feel at the moment is this beauty.

May Sinclair, *A Journal of Impressions in Belgium*, 1915

During the First World War, many women with no previous clinical experience found themselves immersed in medical work in Britain and overseas. From the Western Front alone, around 1¼ million casualties were sent home for treatment.[1] Nursing was the most obvious female niche, the one that confirmed expectations of women as ministering angels dressed in impeccable white as they soothed the fevered brows of the wounded. During parliamentary debates about the vote, one MP singled out nurses for special mention: 'Whose death was it that sounded like a clarion cry throughout the Empire more than any other? It was the death of a woman—Miss [Edith] Cavell. Who have sacrificed their lives in the same way as men through fever, in the wards of hospitals and elsewhere?'[2] The other iconic heroine is Vera Brittain, the heartbroken student who lost her fiancé and relinquished her life at Oxford to care for wounded soldiers.

Figure 13.1. Marie Stopes in her laboratory, 1904.

Despite being so celebrated, Cavell and Brittain made less impact on the nation's welfare than an unconventional scientist who challenged all the stereotypes: Marie Stopes (Figure 13.1).[3] Three decades before the National Health Service was established, she disseminated basic knowledge about sexual activity—euphemistically known as 'moral hygiene'—for those either too embarrassed or too poor to consult a doctor. Already internationally renowned among geological specialists for her pioneering research into plant fossils and her attempts to improve coal production, she became a household name after 1918, when her book *Married Love* was first published. One of the first bestsellers ever, 17,000 copies were bought in its first year, and by 1955 well over a million copies had been printed.

According to Stopes, the immediate inspiration for writing *Married Love* came from a female medical student at Manchester, where Stopes lectured on palaeontology. Because of the War, there were twice as many female students in Manchester as previously, many of whom had switched into vocational subjects such as medicine or engineering.[4] The young woman told Stopes about an incident she had witnessed when volunteering at the local hospital. A tearful mother had brought in a sickly baby who was clearly destined to die, just as his three older siblings had done. Although the male doctor assured the mother that nothing was wrong with her husband and that she should continue to bear his children, it was obvious to the student that the infant had inherited syphilis.[5]

Stopes structured her autobiography as a series of conversion moments, a narrative device adopted by many medical women.[6] According to her, when she heard about the syphilitic baby doomed to die, she realized in a flash that whereas privileged women had known about protection and birth control for years, others were effectively having disease and motherhood forced upon them. Stopes became increasingly convinced that she had a higher calling. In the summer of 1914, she experienced another spiritual awakening, this time on a Northumbrian beach. Her first marriage had recently collapsed, and she was camping by herself, composing a long poem and unaware in her isolation that the War had started. Suddenly, she 'saw, green and horrible, the corpse of a young man and another standing over him smiting at his dead body. The impression was most vivid....' From then on, she reported, she resolved to expend her energies not on what she wanted to do, but on what she felt was necessary, whether it be coal research or sex education.[7]

Using a military metaphor, Stopes later remarked that 'the main ideas in the book crashed into English society like a bombshell.' During the War, military and medical authorities repeatedly denounced what they interpreted as female promiscuity, yet refused to offer advice other than abstinence. At one military hospital, the officer in charge judged that his role was to protect the national safety, not the national health: 'the less said about sex matters ... the better,' he pontificated.[8] Even though 20 per cent of the soldiers returning to Britain were infected with a sexually transmitted disease, doctors denounced her practical books as advice manuals for prostitution.

Published with her second husband's financial help, *Married Love* was an early example of the popular non-fiction books that proliferated between the Wars. This was a wartime work for the middle classes, written—Stopes declared—in the hope of serving the state at a time when happy homes were needed more than ever. She revealed that 'among my most ardent correspondents are young lieutenants recently married, or about to marry'—and in the similarly fraught period after the Second World War, four further editions appeared.[9] From her opening sentence—'Every heart desires a mate'—Stopes stressed the romance of marriage and the rights of wives, adopting the flowery language used by Edwardian novelists as she pleaded for reciprocal sexual fulfilment. She wanted, she claimed, to prevent other brides from her own experience of a marriage that had collapsed because of her sexual innocence (her ex-husband's version of events emphasized her relentless avidity).

Stopes carefully ensured that her book appeared academically valid: she reproduced graphs of female desire based on her own scientific research, listed all her scientific qualifications on the title page, and included professional endorsements. Although doctors deplored her reliance on personal testimony rather than physiological expertise, most of her readers assumed that she was medically trained. Her advice changed the lives of many thousands of men and women who had physically survived the War but who were suffering emotionally.[10]

Scientific Medicine

The urgency of setting up medical relief provided an ideal opening for rich wives who excelled at organizing—professional committee ladies, as it were. Historians still find it tempting to caricature such charity workers as 'titled ladies . . . formidable women who were used to being listened to and treated with great respect, and who usually got their way'.[11] Yet this forcefulness and willingness to cut through bureaucratic red tape could prove crucial. Lady Leila Paget—a diplomat's wife—had already helped to run a military hospital in Belgrade, but as soon as war was declared she rushed back from holiday in California to recruit volunteers and purchase supplies. Leading a team of four doctors and sixteen nurses, she sailed out to Skopje (then known by its Turkish name of Uskub), where she took over a converted secondary school with 330 beds. Despite her attempts to ward off lice by wrapping herself with protective bandages soaked in petroleum jelly, she nearly died of typhus, and as fighting intensified, she was eventually forced to flee in 1916. Like other British women who carried out medical work on the Eastern Front, Paget was decorated by the Serbian government, and her name was added to the saints and queens listed on the wall in a local church.[12]

For other wealthy landed ladies, the War provided an unprecedented opportunity to exercise their brains instead of their dogs and to find some sort of purpose in life. Abandoning the multiple names she had inherited from her royal ancestors, Helena Gleichen threw herself into medical work. Surprisingly, her extraordinary published memoir seems to have been ignored by historians. As the cousin of George V, she had dined with Queen Victoria, danced at debutante balls, and spent much of her life on top of a horse. But in 1915, after working for a few months in a castle that she had borrowed and converted into a hospital, she trained for six months as a radiographer. Her family paid for a portable X-ray unit powered by a converted Austin, and she was all set to work at the front—until she encountered a bureaucratic obstacle. Whereas in France, the government was welcoming the fleet of X-ray cars organized by Marie Curie and her daughter Irène, in Britain the War Office informed Gleichen with unassailable illogic that because no women were radiographers, therefore it was impossible to employ her.

Undeterred, Gleichen and her friend Nina Hollings drove to Italy, where they took many thousands of X-rays, often at risk from enemy bullets and eventually damaging their hands and eyes. Weighing down their un-anaesthetized patients with sandbags, they explored precisely

Figure 13.2. Helena Gleichen's sketch of locating a bullet.

yet rapidly to locate bullets lodged in a soldier's body by a rough-and-ready technique of aiming along two perpendicular lines to see where they intersected (see Figure 13.2). Alone in the middle of a battle zone, the pair learnt how to service their electrical equipment and fix their car—on one occasion, Gleichen enterprisingly melted two large altar candles to replace some cracked insulation. Their photos show two demure women in heavy uniform, hats firmly in place even when working inside, yet Gleichen recounts adventures as hair-raising as those of any front-line soldier.[13]

Like radiography, physiotherapy (which included electrical treatment) was a relatively new medical science that flourished during the War. Rehabilitation therapy was initially funded privately by a philanthropic family who converted their central London house into an outpatient clinic, treating 200 men a day. Their electrical devices, designed to stimulate muscle movement and relieve pain, were later adopted at Army convalescent centres throughout the country. The Honorary Major in charge of the Military Massage Service was an independently wealthy woman doctor—Florence (known as Barrie) Lambert, who sat on the War Office's Electro-Medical Committee. Great care was taken to monitor these intimate encounters between therapists and patients. Depending on the location of a wound, massage could be embarrassing but was above all painful—and humiliating when soldiers were reduced to screaming or weeping. Advertisements for courses promised to protect trainees against inadvertently landing up in 'undesirable houses', and the women's uniform was closely specified: the skirt to be six inches above the ground, the buttons to be bound in navy-blue leather (imitation permitted), jacket pockets to be 8 by 5½ inches.[14]

Women also dominated another medical field stimulated by the War—psychotherapy. Unable to obtain conventional posts for treating the mentally ill, female doctors gravitated towards London's Medico-Psychological Clinic, which operated from 1913 to 1920 and had a permanent effect on the British medical profession's approach to Sigmund Freud. Initially, it was based at the home of an arts graduate, Julia Turner, and her suffragist lover Jessie Murray, a fully qualified doctor who wrote a preface to Stopes's *Married Love* stressing the psychological importance for children of a harmonious family.[15] The Army was beginning to acknowledge the validity of shell shock as a diagnosis, and after a successful appeal for donations, the clinic expanded to experiment with various approaches, including electrotherapy, exercise, occupational therapy, Freudian psychoanalysis, and an eclectic mix of psychotherapies, all offered at affordable rates. By the end of the War, the clientele had grown to 189 outpatients in addition to thirty-six housed in a small residential unit, many of them traumatized soldiers but also 'Non-combatants suffering from War-panic, overstrain, insomnia, and every form of nerve-disaster'.[16]

Female scientists generally chose to enter medical fields along more conventional routes. Hettie Chambers, for example, had graduated in botany at Royal Holloway College and won a research studentship to study green mould. After her husband was killed in 1915, she decided that her skills in high-power microscopy could be put to better use, and she embarked on a hastily constructed course in microbiology at King's College Medical School. Armed with this new skill, Chambers worked at the Seamen's Hospital in Greenwich, but subsequently went back to Birkbeck College and resumed her former life (although in the absence of antibiotics, she died from an infection picked up in the laboratory).[17] Perhaps following the initiatives of Lise Meitner and

Marie Curie, some physicists volunteered as radiologists. Jesse Slater, for instance, was a Cambridge expert on radioactivity who obtained leave to work as a part-time nurse, and then became a full-time radiographer at British military hospitals in France. After the War, she resumed her academic position at Newnham.[18]

Naturally, the War encouraged research into specialities such as surgery, orthopaedics, and tropical diseases. Less obviously, it also stimulated the study of nutrition, which only acquired the formal label of 'food science' during the Second World War. The plant scientist Marion Delf put on one side her research into seaweed and evergreens, leaving Cambridge in 1916 to study vitamins at London's Lister Institute for Preventive Medicine alongside about forty other female researchers.[19] Four years later, she went to South Africa, where she continued to investigate nutrition, influentially demonstrating that the health of mineworkers could be substantially improved by making sure their diet included vitamin C. Another woman at the Lister, Harriette Chick, devised a diet including dried eggs and Marmite, which was used to treat soldiers in Greece suffering from the deficiency disease beriberi.[20]

Some scientists were permanently diverted into applied fields. In the absence of adequate refrigeration, it was difficult to make sure that food supplies were fresh. One particularly prevalent problem was 'ropiness', a common bacterial infection that would cause a loaf to rot and blacken until it could be pulled out in semi-liquid strings. To tackle this, the Medical Research Committee (later Council) employed bacteriologists such as Dorothy Lloyd, who had gained a first-class degree in biochemistry (in 1910) and was only the third fellow to be appointed at Newnham. Dropping her research into muscles and osmosis (fluid transfer), she began investigating the bacilli causing ropiness. After the War, determined to show that women were perfectly capable scientists, Lloyd accepted an invitation to carry out industrial research, and introduced important new techniques for tanning leather.[21]

The War inspired many younger women to become doctors. In 1988, as an elderly woman, Ruth Dingley explained that she had switched into medicine at Newnham in response to the distress of her friends whose brothers had been killed in action—we all felt it 'so *personally*', she emphasized. In 1916, in only her second year as a student, she volunteered to spend a week watching operations on wounded soldiers who were being treated in a temporary surgical hut on a Cambridge playing field. To her alarm, despite several fainting episodes, she was then put in charge of administering the chloroform (better too much than too little was the standard advice). Surviving that harsh initiation, after the War she qualified as a doctor at University College Hospital in London, but not before surmounting further obstacles: greatly outnumbered by men, Dingley and the other female trainees felt themselves to be 'hated like poison' by the hospital staff, including the nurses.[22]

Most medical schools refused to admit women. A woman ran the risk—it was maintained—of becoming masculinized by adopting a professional career. Many people still agreed with the eminent surgeon who had protested in 1879 that 'the acquisition of anatomical and physiological knowledge [represented] the gratification of a prurient and morbid curiosity and thirst after forbidden information—and in the performance of routine medical and surgical duties the assumption of offices which Nature intended entirely for the sterner sex.'[23] Families raised similar objections. 'Don't tell people you are at college,' one elder brother told a medical student; 'and for heaven's sake, don't tell the chaps that you are taking your cases in midwifery. There is something indecent about a girl studying in midwifery.'[24] Resistance was still rife in 1928, when a senior doctor at a London hospital argued in *The Times* that 'the opportunities for athletics must also be less attractive in mixed schools'—to say nothing of the 'inevitable distractions' caused by pretty young women. And then, just as in the universities, there was the embarrassing reality that women often outperformed the men in examinations.[25] Conversely, one of the strongest arguments in favour of allowing women to train was the impropriety of male doctors treating female patients. In 1904, Dr Mary Murdoch declared to her students that she dreamt of a future in which it would be seen as 'one of the barbarisms of a past age that a medical man should ever have attended a woman'.[26]

Campaigning for women's rights did not necessarily entail jettison-ing conventional gender roles. 'I am neither a doctor nor a nurse,' declared the redoubtable Mabel St Clair Stobart, who before the War had led her female troops across Europe to set up a military hospital in Bulgaria, 'but I stand for the principle that all work connected with the sick and wounded is woman's work.'[27] After visiting a military hos-pital in France staffed by British women, the suffragist Evelyn Sharp reflected on

> the bitter irony of our civilisation, which first compels men to tear one another to pieces like wild beasts for no personal reason, and then applies all its arts to patching them up in order to let them do it all over again... somehow, when the patching is done by women the ironic tragedy of the whole thing seems more evident.[28]

Photographs and advertisements portrayed female doctors as ladylike attendants lacking contact with the sordid aspects of medical reality. But even long after they had demonstrated their skill and their stam-ina, they were often regarded with suspicion. Like many of her col-leagues, Dingley found herself in an anomalous position. As a woman who studied medicine, she was conforming to stereotype by entering a caring profession, yet at the same time she was flouting convention by training to be a doctor. This dilemma continued to confront intelligent women throughout the twentieth century. Schoolgirls were increasingly encouraged to study scientific subjects, yet they were still being subjected to social pressures that they should *also* look glamorous *and* be perfect wives and mothers.

Professional Growth

There are several competitors for the accolade of 'first British woman doctor'. The honour is usually awarded to Elizabeth Garrett, who in 1865 received a licence to practise medicine from the Society of Apothecaries (after marrying in 1871, she became Elizabeth Garrett

Anderson). But half a century earlier, Margaret Bulkley—who lived as James Barry—had graduated in medicine and pursued a career as a male military surgeon. And in 1849, Bristol-born Elizabeth Blackwell became the first woman to gain a medical degree in the United States. And what about Sophia Jex-Blake and her colleagues, who success-fully persuaded Edinburgh University to admit women for the first time in 1869? Or Mary Scharlieb, a lawyer's wife who studied medicine in Madras, where she set up a hospital treating Indians as well as Europeans? She was not only the first woman to gain her MD (the top-level medical degree) at London University, but also the first woman to be on the visiting staff of a major hospital.[29]

During the nineteenth century, these pioneering individuals man-aged to find some support through the women-only hospitals and medical schools. As well as caring for the sick, they tried to rectify the impression purveyed by many male doctors that women were physio-logically weak and prone to illness. Writing in 1897, Dr Jane Walker felt she needed to declare that

> A healthy normal woman should have as much enjoyment in the mere fact of living...as a healthy normal man; she should be no less able to make any reasonable physical or mental effort without undue reaction, and her balance of well-being should be scarcely less stable.

The War did at least give women the opportunity to demonstrate the validity of Walker's claim, and she remained an active campaigner for female doctors to be recognized as fully professional on a par with men.[30]

The first women's hospital in Britain dates back to 1866, when the newly qualified Garrett opened a London dispensary for the poor. Six years later, she had raised enough money to install ten beds on the upper floor, and her New Hospital for Women (later renamed after her) soon expanded. Other women's hospitals followed, and by the beginning of the War there were sixteen, including one that happened to open on 4 August 1914 and a couple that also accepted male patients. Supporters insisted that these hospitals were beneficial not only for ill women, but also for the training and career prospects of female doctors. Although the London School of Medicine for Women was set up in 1874, it was initially organized separately from the teaching hospitals, which made it almost impossible for its graduates to obtain house-posts in mixed hospitals.

The second generation of women doctors, those who qualified around the turn of the century, was far larger. In 1891, there were only twenty-five women on Britain's Medical Register, but two decades later, their numbers had risen to around 1,000—roughly 2 per cent of the nation's total. In contrast with Germany, Russia, and the United States, there were hardly any female dentists practising in 1914: twelve, to be precise.[31] Discrimination was rampant. Looking back half a century later, a Scottish doctor wondered how she had found the courage to endure her medical training surrounded by sceptical men who were often more interested in her appearance than her ability. Segregated for dissections, anatomy classes, and ward rounds, even after qualifying these women faced an uncertain future—and they knew that if they married, they would have to resign.[32] As a journalist commented bitterly, female doctors generally had only two options: to be 'sent as a missionary physician to the heathen or limited to a practice exclusively among women and children'.[33]

For talented female doctors, especially those not particularly interested in obstetrics, the best option was emigration to commonwealth countries where they could practise the full range of medicine. Elizabeth Macdonald was determined to become a doctor, but as one of nine children in a poor but enlightened family, she relied on bursaries for her education. Spurred on to do well in her examinations by the prospect of a £100 award, she discovered that 'When the results were posted up there was only one name in the list for First Class Honours.' Her first thought was 'I had won the Scholarship! Joy unbelievable!' The next day she hit reality—as a woman, she was ineligible. By working in a chemist's shop, she managed to complete the course with distinction, but was banned from holding a hospital appointment. Little wonder that she emigrated to New Zealand three years later.[34]

As in other scientific professions, women were paid less than men for doing the same work, were passed over for promotion, and were excluded from medical societies.[35] Access to training was heavily restricted. Cost was merely one obstacle: when the War started, neither Oxford nor Cambridge offered women medical degrees, in London their only hospital option was the Royal Free, and in the other English universities they were excluded from mixed classes on anatomy and dissection. Opportunities were so limited that many women trained in Europe: in 1914, one campaigner protested 'the majority of medical women working for their country to-day have been forced to gain their knowledge and skills in the schools of the enemy.'[36]

During the War, under the double pressure of wounded soldiers and absent men, traditional medical schools were forced to reconsider their policy and admit women. But even so, four of London's major hospitals remained resolutely all-male, and after the War, some medical schools once again closed their doors to women. Although there were over 2,000 female doctors by 1921, these women who had invested their studies and their savings to secure a viable career were repeatedly thwarted: it was hard to gain promotion in a profession dominated by resentful men protecting their status. In 1917, Ida Mann experienced a foretaste of such resistance when she was put in charge of an antagonistic physiology class containing male medical students who had been recalled from the front to complete their training. 'Damn and blast you,' one of them swore at her; 'five days ago I was killing Germans. How the hell can you expect me to spend the afternoon tying little bits of cotton and wire to a dead frog?'[37]

The War gave qualified doctors unprecedented opportunities to work both at home and abroad. Most obviously, as men went off to the front, vacancies arose in civilian hospitals and practices, where the women's red shoulder straps identified them as doctors.[38] New wartime positions opened up, so that female doctors were recruited to monitor health in munitions factories, inspect food processing plants, or run the maternal advice centres that were multiplying to nurture children for safeguarding the nation's future. For the first time, qualified women doctors were in demand, and mature students as well as schoolgirls were being encouraged into medicine. In 1914 only 700, or 10 per cent, of medical students were female, but by 1918 the number had more than tripled to reach 40 per cent.

Increasingly, female doctors became involved in military medicine, but because they had mostly worked in hospitals for women and children, they had very little experience of treating men's bodies, let alone those of casualties with traumatic war injuries. Time after time, military patients said that once they had got over the initial surprise, they appreciated being cared for entirely by women. Whether or not they supported suffrage, these soldiers valued these doctors' feminine, nurturing behaviour. A soldier's letter from France was reproduced in the national press: 'I got hit by a shell bursting over our trench—in the face, neck and shoulder. I am in one of the very best of hospitals—a ladies' hospital. Lady doctors do all the work—no men at all, so you can guess I am all right.'[39] When a ward of Belgian soldiers presented Stobart's all-female team with medals and flowers, she was overwhelmed with emotion—her only regret, she wrote in a women's magazine, was that a certain 'eminent English surgeon, with his mid-Victorian notions as to the sphere of work of women, could not have been present at that little ceremony'.[40]

The first woman doctor to work for the War Office in England was Florence Stoney. Although she had plenty of experience in a relatively new medical field—she had set up the radiology facilities at the Royal Free Hospital and the London School of Medicine for Women—the War Office initially declined her offer of help. Like other specialists in her position, she joined one of the women's voluntary organizations and travelled to Europe, gaining an experience unusual even for men— dealing with wartime fractures that had gone septic after remaining untreated for several days. Her sister Edith, who taught physics at the London School of Medicine for Women, helped to set up their unit's portable lighting system, which was essential in the absence of a regular electricity supply. This could be hard physical work, with heavy equipment to carry, and emergency repairs during night-time storms.

Edith Stoney acquired plenty of electrical and medical expertise, but she claimed that even after three years' arduous experience, she had no suitable position because the British War Office would not let female radiologists or wireless operators work abroad. A Scottish doctor serving in Serbia remarked admiringly that the physicist was thin but had stamina: dressed in 'rubber overall, thick rubber gauntlet gloves, spectacles on the end of her nose, her fair hair streaked with grey gathered tightly up into a top-knot on the summit of her head, her whole mind centred on her work'.[41] In a letter soliciting employment, Edith Stoney boasted about wiring machinery together in sub-zero temperatures, and providing illumination for female surgeons in France: '50 candlepower were lamps I had stuck in cocoa tins for shelter—we dared show no light of course—but had to run the engines for the X-rays.' Perhaps, she speculated, 'the British would let me go on a hospital ship ... I should be setting a skilled man free, and the danger is not such as a woman cannot face.'[42] But although women judged her to be brilliant, she could be hard to get on with, and had the reputation of 'grumbling, and nothing being good enough—I think it is rather a pity that she has been taken on the staff at all.'[43]

In 1915, the War Office decided to bring casualties back home for treatment—and it also (at last) took notice of all the glowing reports on women's hospitals. Florence Stoney was asked to take over the X-ray department at the large Fulham Military Hospital, and through her sister Edith she recruited science students who volunteered to work the equipment. The smart ones soon found themselves in the operating theatre, digging out pieces of shrapnel and sewing up wounds. Although Stoney and her female assistants examined over 15,000 injured soldiers, as a civilian doctor employed by the Army she was given only a short-term contract and paid on a daily rate. However distinguished they might be, these women were denied a rank, any possibility of promotion, and the all-important official military uniform.[44]

Figure 13.3. Louisa Garret Anderson (*left*) and Flora Murray (*right*).

Two doctors in particular became famous in wartime Britain—the surgeon Louisa Garrett Anderson (daughter of Elizabeth and niece of Millicent Fawcett) and her long-term companion, the physician Flora Murray, both of them committed to improving the professional status of women (Figure 13.3). Women were setting up voluntary hospitals all over the country, but Anderson and Murray were different.[45] Inspired not only by patriotism but also by their suffragist ideals, they refused to employ men and flouted War Office advice by travelling abroad and converting a central Paris hotel into a hospital staffed entirely by women.

This venture proved so successful that within only a few months military officials asked Garrett and Murray first to organize a second French hospital, and then to come back home and run a large Military

Hospital with over 500 beds in an old London workhouse, where they treated around 26,000 patients (including 2,000 women) before it closed in 1919.[46] Set in Bloomsbury, close to the railway stations bringing in fresh convoys of wounded soldiers, the Endell Street Hospital was an extraordinary institution. Financially supported by the Royal Army Medical Corps, it was run by women hailing from the Dominions, the United States of America, and all over Britain. In conventional military hospitals, officers were cared for by four times as many nurses as ordinary soldiers, while the specially privileged were pampered by volunteers in the private homes of wealthy Londoners.[47] But in Endell Street, treatment was equal. Officers and privates lay side by side in the same wards to cry, laugh, and suffer together. The corridors were often packed with anxious relatives trying to interpret the expressions on a doctor's face. Murray remembered terrified elderly parents from Ireland and Scotland on their first long trip, who 'would not venture in the streets in case they should be lost or run over.... The relatives were very brave and very pathetic.'[48] Thoughtfully, Anderson and Murray limited visiting hours to protect their patients from being overwhelmed by well-intentioned ladies indulging a voyeuristic fascination with mutilation.

Perhaps oddest of all, the Endell Street Hospital was an overtly political organization, publicly celebrated in the press as a suffrage hospital; even its official motto—'Deeds not Words'—was borrowed from the suffragette movement, which also gave them money. Before the War, most female doctors had been so anxious to preserve their professional reputation that they refrained from open political campaigning, instead supporting feminist causes unobtrusively by donating money, providing first aid to women injured in demonstrations, or joining the Women's Tax Resistance League (when the Inland Revenue seized possessions in lieu of tax, League members simply bought them back again).[49] In contrast, both Anderson and Murray had been militant suffragettes and remained active campaigners: a photograph taken in Paris shows their purple, white, and green buttons pinned to the jackets of the uniform chosen by Murray—khaki with a dull red piping.

Anderson and Murray wanted to prove that women could work as professionally as men—but they also insisted that women's medicine would be different. In her scrapbook, Murray carefully preserved a cutting from the *Tatler*, which praised 'the noble ladies who manage the Suffragette hospital in Endell Street. They are men in the best sense of that word, and yet women in the best sense of that word also.'[50] Somehow, they found time to organize weekly educational sessions, encouraging the staff to become good citizens by teaching them about famous women such as the nineteenth-century scientist Mary Somerville.

Whereas previously the two doctors had restored Cat and Mouse suffragettes until they were fit enough to be recaptured, now they were engaged in the still more heartbreaking task of curing wounded soldiers so that they could be sent back to fight and perhaps be killed. A *Punch* cartoon of 1915 caricatured Anderson as an 'eminent woman surgeon' talking to an injured guardsman in a hospital bed, wondering where she has seen him before. 'Well, Mum,' he replies, 'bygones be bygones. I *was* a police constable' (see Figure 13.4). Similarly, when Sylvia Pankhurst visited Murray, she remembered being treated by her in prison. Observing her from a distance, Pankhurst realized that although she appeared flat-chested and emaciated, Murray embodied the tireless 'efficient activity ... of the intelligent middle class woman, whom an earlier generation would have dismissed as an "old maid"'.[51]

Female doctors faced resentment not only from army officers and medical men, but also from patients worried about what was happening to their wounded bodies. 'When she was doing me I could see it was different to what I had,' one man complained. 'She says, "Don't you tell me how to do it," so I had to stick it .... What the devil did she know about it? She is not going to practice [*sic*] on me (not likely). I have stuck enough.'[52] At Endell Street, the female staff 'had this drilled into us: you not only have to do a good job but you have to do a superior job. What would be accepted from a man will not be accepted from a woman. You have got to do better.'[53] Even the most fiercely independent of women could relish the discipline of a military hospital,

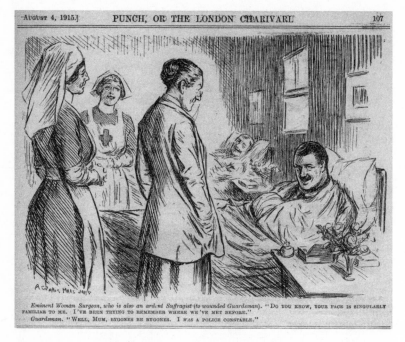

Figure 13.4. Arthur Wallis Mills, cartoon in *Punch*, 4 August 1915.

feeling 'that my irrelevant former self, with all that it has desired or done, must henceforth cease (perhaps irrevocably) to exist.... From the moment that the doors have closed behind you, you are in another world, and under its strange impact you are given new senses and a new soul.'[54]

More used to paediatrics and obstetrics than trauma surgery, Anderson and Murray were coping with lacerated abdomens, gas gangrene, broken limbs, and injured skulls. Unsurprisingly, their relationships with Army personnel were tense ('Good God! *Women*,' exclaimed one disgusted colonel, 'what difficulties you will have'), so they used women's medical and suffrage networks to recruit volunteers and sympathetic staff.[55] Among the dozen or so doctors, Helen Chambers had previously been a lecturer at the London School of Medicine for Women, but she temporarily put to one side her research into cancer and became their consultant pathologist. An expert on radium therapy, still in its infancy, Chambers later played a key role in international cancer trials and contributed to the relatively high survival rate at Hampstead's Marie Curie hospital; like so many of these early radiologists, including Curie, she herself died from cancer. Collaborating with Anderson, Chambers helped to devise and test new treatments, notably a money-saving antiseptic paste that reduced the frequency with which surgical bandages had to be changed, which was a painful process. In addition, the team invaded yet another predominantly male domain by publishing seven papers on war-related injuries in the *Lancet*.[56]

Medical Women and Female Doctors

In 1915, waiting to interview a female doctor, a journalist listened in on a conversation.

> In a corner apart sat two patients talking volubly; as they grew more interested they raised their voices. Said one of them: 'Yes, I always come to her in spite of the distance: I like her. She is such a womanly woman, is she not?' to which the other replied: 'Yes, I suppose so, but what *I* like about her is that she really gets me well quickly.'[57]

These two unnamed women were articulating a fundamental dilemma that plagued many female doctors: as professionals, were they equal to men or were they fundamentally different? Or was there, perhaps, some way that they could be both?

Anderson and Murray had been militant suffragettes, and they antagonized many male medical colleagues by their unwomanly choice of career. Yet in other ways, they fulfilled conventional expectations of behaviour. Naming their wards after female saints to replace the military-style letters of the alphabet, they brightened the rooms with fresh flowers and coloured blankets, and cared for their patients psychologically by laying on theatrical performances, sports events, and a large library. In an opening speech, Anderson underlined her maternal approach:

> After all, if you have found out the way to treat children—what toys they like, what they like for tea, and what frightens them when going to an operation—you have gone a great way to find out how to run a military hospital [with] 550 large babies requiring a great deal of care, a great deal of understanding, and a certain amount of treatment.[58]

Yet despite all these feminine touches, Anderson and Murray insisted that theirs was a military hospital just like any other. Determined to appear totally professional, they adopted a military-style uniform with RAMC badges on their lapels, replaced their WPSU badges with campaign ribbons, and refused to let Agnes Conway include Endell Street in the Imperial War Museum's collection of women's war work. After the War, the Museum commissioned an artist to visit the hospital. Anderson and Murray were appalled by his picture, complaining that it included features that should never have been (but apparently were) in the operating theatre—sterilizing drums, a kneeling nurse (Figure 13.5). At their insistence, a different artist produced the official oil painting that is now displayed, a heroic but misleading vision of a calm team working in an orderly room.

In the intimate surroundings of a hospital ward, gender stereotypes were challenged. Brought down from their heroic role as brave soldiers defending the nation, male patients were weak and dependent not only on feminine nurturing but also on women's physical strength. Prescribed occupational therapy, they found themselves carrying out conventionally female tasks such as knitting, embroidery, or making woolly rabbits. Turning to each other for moral support, some wounded men developed close and caring relationships that were not necessarily sexual, but that conformed to feminine expectations of behaviour. These male patients resented being under the control of women who had the right not only to hurt them but also—like their military superiors—to discipline them for breaking petty regulations or pretending to be more sick than they really were. Women had not

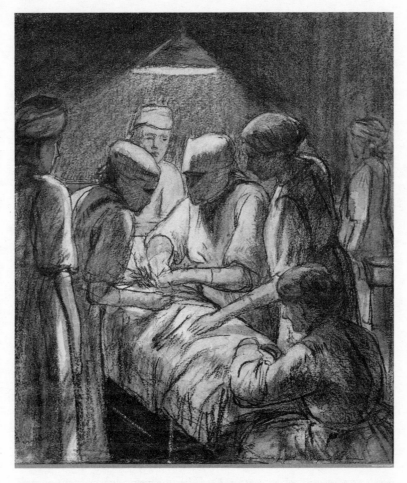

Figure 13.5. An operation for appendicitis at the Military Hospital (detail), Endell Street, London. Chalk drawing by Francis Dodd, 1917.

only taken over their peacetime jobs in factories and farms, but were also caring for them as if they were children. While the doctors might be held up as mother figures, physiotherapists were cast as torturers dispensing physical pain, and nurses were converted into guardian angels and romanticized as lovers.[59]

When the House of Lords gave women over thirty the vote in 1918, there was great excitement at Endell Street. The courtyard was decorated with flags, the doctors wore their academic gowns over their uniforms during dinner, and the staff joined a large crowd singing the suffragette anthem, 'March of the Women'.[60] But although female doctors rejoiced, for many of them it seemed still more important that they be treated as fully fledged professionals. A few exceptional individuals pursued illustrious careers after the War, but many were forced back into looking after women and children, no longer deemed suitable as surgeons or specialists. Ironically, the ones who suffered most were those who had benefited from the greater training opportunities during the War but still lacked substantial clinical experience—like most employers, hospitals gave priority to their male contemporaries.

Yet the War had enabled women to demonstrate their effectiveness as doctors. In 1917, the Medical Women's Federation was founded in order to campaign for full recognition and equal terms of employment. Throughout the interwar years, it demanded several rights—to continue practising after marriage, to retire at the same age as men, to receive adequate pensions. But above all, these doctors insisted on equal pay for equal work, arguing that accepting lower salaries would result in lower prestige—an important principle, and still an outstanding problem.

Naturally, there were internal conflicts: just because they were all women did not mean that these doctors had identical points of view. Whereas younger members regarded gender as irrelevant—technical competence was what mattered—others maintained that they had a dual role as both women *and* doctors. In 1929, Dr Ethel Bentham wondered if there was 'still useful work for separate organizations—political or professional—of men and women?'. That question remains unresolved. Should women fight for their rights separately or in collaboration with men?[61]

14

FROM SCOTLAND
TO SEBASTOPOL

The Wartime Work of Dr Isabel Emslie Hutton

The ladies' cloakroom, with its pavement, its hot-water supply and basins, was converted into the operating theatre. A gas steriliser and a powerful electric light were fixed in. Though small, it proved service-able and well equipped....Late in the afternoon a message from the Chief Surgeon was brought from the wards: 'Dr Anderson wishes to know if the theatre is ready and if she can operate tonight.' 'It is ready now' was the answer. And the surgeons operated till a late hour that night.

Dr Flora Murray, *Women as Army Surgeons*, 1920

'My good lady, go home and sit still.' That was the War Office's curt response when Dr Elsie Inglis offered her services as a surgeon in August 1914. Furious at this official rejection, she appealed for help first to Scottish suffragists and then to women's groups all over the country, soon raising enough money for two fully equipped hospital units, staffed entirely by women. Although spurned by Brit-ain, the services of the Scottish Women's Hospitals organization were gratefully accepted by two allies, France and Serbia, where they saved many thousands of lives.[1] Towards the end of the War, one Scottish doctor smiled smugly to herself when a British Army colonel in Serbia was so short of medical help that he asked her hospital to be officially responsible for his drivers.[2]

Inglis's story is as gripping as a novel. Born in India in 1864, she was among the earliest group of Edinburgh doctors to qualify and also a leading figure in the Scottish suffrage movement. Already suffering from cancer at the beginning of the War, she went out to Serbia and worked under atrocious conditions caring for wounded soldiers until she was captured, imprisoned, and repatriated (see Figure 14.1).

The situation in Serbia was horrific. A former suffragette described her

> nightmare of a day—one ward just straw—men laid down without taking clothes off. Terrible sights and sounds—lots of very bad cases— gangrene—smell at times almost unbearable—a strange sickly odour quite unlike anything else, which seems to permeate everything and stick in one's nostrils, even when one is out in the fresh air.[3]

Some doctors resorted to taking morphine, which in powdered form was relatively easy to steal in small amounts; if troubled by their consciences, they could always argue that as addicts, they were as sick as their patients.[4] Undeterred by her prison experiences, Inglis led a fresh team to the Eastern Front, but illness forced her back to England, where she died in a station hotel the day after landing.

Hardly surprising, then, that Inglis has become a wartime heroine, feted for demonstrating that female initiative and persistence could overcome military obstinacy. But other aspects present a less saintly image. One of her staff noted that on the ship to Odessa, 'Dr Inglis has been rather bitten with the craze for playing at soldiers and likes people to salute & stand to attention etc etc.' Similar reports confirm her view that 'Dr Inglis is not a good leader. She gets very much fussed & loses her temper easily, rapidly & often unjustly.... [S]he takes no personal interest in anyone and is therefore not at all popular.'[5]

Exceptional as Inglis was, other doctors led equally extraordinary lives. Dr Elizabeth Ross, for example, is scarcely visible in Britain's official records, but every year on St Valentine's Day her death is commemorated in ceremonies throughout Serbia, where her name has become a symbol of courage. After qualifying in Glasgow in 1901, she spent several years working in Persia (now Iran) as the assistant to an Armenian physician: her proud descendants still treasure a photograph of Ross wearing traditional Bakhtiaran costume and smoking a

Figure 14.1. Elsie Inglis and some of her medical sisters, 1916.

hookah. Next, after applying through a newspaper advertisement, she travelled along the coasts of Japan and China as a ship's surgeon: she was probably the first woman to be employed in that position. But as soon as Ross heard about Inglis and the Scottish hospital units, she borrowed some money and went to join them in a large Serbian military hospital. Only three weeks later, she died of typhus, infected by her own patients.[6]

Among the many similar medical casualties, Dr Laura Forster had initially volunteered to work in Belgium, but then decided she could be of more use in the men's surgical department of the largest hospital in Petrograd (St Petersburg). Unlike Britain, Russia had conscripted all its doctors—female as well as male—so there was a desperate need for professional medical care. Forster was later recruited to join an all-female team organized by the NUWSS in a Russian maternity hospital, but as the War intensified, they were sent behind the lines to treat cholera, smallpox, and other infectious diseases. Along the way, they stopped at a goods station converted into a temporary hospital, where 9,000 wounded soldiers poured in during the second night alone. A year later, in February 1917, Forster's death was attributed to 'heart failure after influenza'.[7]

In contrast, Dr Caroline Twigge Matthews stared at death several times but repeatedly escaped. Determined to match the wartime contributions of men, she set off independently on a converted tourist liner newly fitted out with naval guns, which was chased by a sub-marine as they drew near to Malta. After several similarly nightmarish adventures, she landed up in a large Serbian military hospital, left behind at her own insistence to care for those too sick to flee from the approaching enemy. Sleeping on bare boards, constantly hungry, she soon learnt to ignore the swarming rats and cockroaches as she looked after her abandoned patients. Once the invading Austrians had arrived, she struggled on for weeks, effectively a prisoner of war and working right through a bout of diphtheria. Accused of espion-age, for several days 'Twiggie' was hauled across icy mountain ranges in a bullock cart, destined for execution in Belgrade but eventually arriving in a Hungarian prison where—if this were a novel, it would be too corny to believe—her fellow captives included Alice Hutchison, an older (and much admired) Edinburgh doctor whom she had first met on the voyage to Malta. Matthews's book is packed with jingoistic assertions of British superiority, yet despite the ordeal she patriotically endured at her own expense, she has no entry in the *New Oxford Dictionary of National Biography*.[8]

When the War started, the British Army was adamant: female doctors should stay at home to fill the vacancies created by men going overseas. 'I cannot think the front, or very near it, is the place for them,' declared an influential committee chairman. Within a few months, women had taken the initiative into their own hands, setting up hospital units all over Europe. Headed by the Liverpool surgeon Frances Ivens, an Inglis team that included three surgeons, two physicians, and a radiologist was promptly despatched to a town near Paris, where the women took over Royaumont, a disused Cistercian abbey. The patients' beds were lined up in the cloisters, while the doctors and nurses made do with mattresses on the stone floors of monks' cells, trying to ignore the bats that constantly flew in and out.[9] Before long, all-women units had settled still further away in Corsica, Salonika, Russia, Serbia, and other sites along the Eastern Front.

Living and dying on the edge of danger, these doctors had an enormous impact on the war zones and their local populations. Most obviously, they rapidly acquired the surgical expertise needed for treating wounded soldiers, and countless affidavits testify to their patients' appreciation. In addition, they ran maternity units, cared for refugees, researched into infectious diseases, and introduced preventive health programmes.[10]

The success of these self-propelled organizations eventually convinced the War Office that female doctors could be effective. They started by employing women doctors in military hospitals at home, but by 1916 demands for help had increased so much that they recruited women to serve overseas. Yet despite constant lobbying by medical societies, and despite equal workloads, the Army continued to treat its male and female doctors differently. Predictably, the women were paid less—but what really rankled was their low status, their civilian clothes, and their inability to claim routine compensation such as adequate travel expenses. In some areas, being denied an identifying uniform was positively dangerous: spies could infiltrate British hospitals, and, conversely, British doctors in civilian clothes ran the risk of being captured if they ventured into nearby towns. 'What a rotten position these women have,' one of Inglis's doctors wrote to her mother, lamenting that 'a decent old soul' in her fifties had 'got no rank & [is] junior to the most junior male doctors'.[11]

In contrast, the women who worked for and with other women had the nearest possible experience of what it felt like to be a man. Some of them had reservations about veering too far towards masculine behaviour. 'This is the strangest affair I have been mixed up in,' reported a new recruit from her ship. 'They wear very military khaki coats with breast pockets, belts etc; the merest shadow of a skirt and puttees. They call each other "chaps" and say "Yes Sir" to Mrs Haverfield.'[12] But even those who refused to renounce their femininity could enjoy responsibility and make decisions about their lives, often for the first time. As they learnt how to cope with extremes of weather and harsh terrains, they were fulfilling their own desire to defend their country as well as to nurture the sick.

The worst conditions were in Eastern Europe, where many women remained throughout the War. 'Our food which we brought with us consists of hard boiled eggs now 8 days old, bread of the same date & so stale that you have to cut it up with a knife & fork,' one correspondent reported miserably in a pencilled note;

> Our candles are almost finished and we have to sit in darkness most of the time.... There are swamps of dirty water everywhere, the mud chronic, and the smells horrible. Dead horses & dogs lie about, heaps of dirty dressings are thrown all over the station which is crowded with hospital trains. When we first arrived it was bitterly cold & snowing.[13]

Some of them died, succumbing either to infectious diseases, the punishing environment, or enemy bullets, but for the survivors Armistice Day passed virtually unnoticed as they struggled to contain the flu pandemic that killed almost 5 per cent of the world's population (see Figure 14.2).

Either too busy or too modest to write about themselves, these women have left few traces of their existence. One exception is Isabel Emslie Hutton, who in the 1920s set out to plug a gap in the war literature by writing about the conflicts in Salonika and the Crimea rather than along the Western Front. By choosing the unusual strategy of focusing on 'the emotions and sensibilities of the gallant souls who suffered alike in victory and defeat', she inevitably revealed much about her own character.[14]

Figure 14.2. View of field-hospital ambulances and tents, Macedonia.

An Edinburgh Doctor

Dance was the thread that Isabel Emslie Hutton wove through the pattern of her life. In Edinburgh, as a child she had admired male performers at the Highland Games, and the night that she gained her doctorate she went to the theatre and marvelled at Anna Pavlova. Temporarily based in Sebastopol just after the War, she taught the Highland Fling to the head of the Bolshoi ballet, and in exchange learnt old Russian court dances that she subsequently performed in Istanbul to raise money for starving Russian refugees stranded on boats in the harbour. Later, living in London, she sat next to Arnold Bennett when Igor Stravinsky conducted his ballet *The Wedding*, and saw Pavlova dance for the last time in public. Long afterwards, she learnt from another doctor that the ballerina suffered intense pain throughout the performance, but had been determined to continue. Hutton was the same sort of person. Taught by her parents to give away all but one of her Christmas presents on New Year's Day, she committed herself to medicine at the age of seventeen, renouncing with the dedication of a nun all thoughts of marriage.[15]

Brought up in a strict Presbyterian family, Isabel Emslie spent much of her Victorian childhood visiting the sick and the poor, experiences that underpinned her determination to become a doctor. Despite their disciplinarian approach, her parents were willing to pay for a daughter's education, and her mother abruptly silenced any friends and family who dared to disparage women doctors. Even so, during her five years of medical school, Emslie and the handful of other female students repeatedly felt their self-confidence ebbing away in the harsh atmosphere. She won prize after prize, but Emslie knew that after graduation, the men were guaranteed secure jobs for life, while she contemplated...what? Life overseas with missionaries? A post in the remote Highlands?

She plumped instead for three years as a pathologist in a Scottish mental hospital, where she carried out research into chemical tests for syphilis. Willie Macdonald, the patient she recruited as an assistant, turned out to be so efficient that they were soon in business selling experimental guinea pigs to laboratories all over Scotland. After she gained her MD, her boss told her that as a woman, she stood no chance of promotion, and waited for her to open the door for him so that he could leave. Instead, Emslie exited not only the room but also the asylum—and started her professional career as a doctor in a children's hospital. After six months, she had made such an impression that she was offered two extraordinary jobs. The more prestigious and the more tempting was an invitation—the first extended to any British doctor—for the Mayo clinic in New York. But she turned that down in favour of going to a small but progressive Edinburgh mental hospital with an unusually high proportion of female staff.

Psychiatry was still a new speciality, and by 1914 Emslie was deep into studying Freudian and Jungian theories. As the War continued through the winter, her life changed, first gradually and then abruptly. Noticing that her patients (both male and female) had begun knitting for the Red Cross, she published an innovative paper in the *Edinburgh Medical Journal* about the effects of war on the insane—one of the few articles on psychiatry at the time, and one of the even fewer written by women. Summarizing her own case notes, she argued that during its first couple of months, the War had excited a small number of illnesses and aggravated others.[16]

Emslie loved her work, but as the men left, she found herself managing a depleted and elderly workforce. The temptation grew to travel overseas. Perhaps, she thought, she might get the opportunity to practise surgery, her favourite speciality but one normally reserved for men. By August 1915, she had made up her mind to join Inglis's Scottish Women's Hospitals. She prepared for her mission carefully. Knowing that she would be in charge of the pathology department in Troyes, one of the two women's hospitals in France, she recruited a member of her laboratory team as an assistant, and set off from Edinburgh with two nurses.

Whoever designed the smart grey and tartan outfit, with its long skirts, embroidered thistles and neat little shoes, had little appreciation of the mud and squalor that lay ahead: the women had discarded their headgear (secured with painful hat-pins) before they even left Paris. Arriving at Troyes, which was funded by the women's colleges at Cambridge, they discovered a long, white château with 250 patients arranged in open-sided tents across the lawns; the orangery made a delightfully light and airy operating theatre. A few weeks later, rumours started circulating about an imminent move, and Emslie became acquainted with the opacity of military intentions. First they were told to pack up the hospital—but that order proved premature. Soon they were putting the tents up again to accommodate 200 soldiers coming straight from the battlefield.

Eventually, in October 1915, they really did leave, but for an unknown destination. It turned out to be Salonika, a Turkish town that had belonged to Greece for the last three years.

Salonika

The British encampment in Salonika seems to have been an embarrassing, expensive mistake. The first contingent arrived in autumn 1915, unfortunately too late to fulfil its mission of protecting the Serbians against the Bulgarians. Within a couple of years, around 400,000 troops were stationed there, most of them suffering from malaria and serving no clear military function. An Irish MP told the House of Commons that conscripted soldiers were crying out '"Anywhere but Salonika"', knowing they would be 'put into a swamp to wait until they are struck down by sickness'. While the government desperately tried to cover up the situation, medical and army officials accused each other of being responsible for failing to stem malaria, still a poorly understood disease. There was little fighting until a couple of months before the Armistice, when an extraordinary attempt to storm a high mountain ridge defended by fortresses proved (unsurprisingly) disastrous.[17]

On arriving in Salonika, Emslie's first health measure was to cut off everybody's hair, starting with her own, which had grown to waist-length. Despatched on an uncomfortable train journey to Ghevgeli, about fifty miles north, the women found their moribund patients crammed together on filthy straw pallets, the surgeons operating by the light of thin tallow candles with neither disinfectant nor anaes-thetics. In the icy conditions, frostbite and gangrene added to the complications of dysentery and other infectious diseases, while every-body's sleep was disturbed by the howls of jackals scavenging for food and corpses. The women built incinerators, dug latrines, erected tents, installed X-ray equipment, and set up a dispensary in a disused silk-worm factory. Soon they had transformed the rickety buildings into a functioning hospital. But only a few weeks later, Serbia was invaded. Once again, they had to pack up the hospital that they had so recently created and move on, this time to a hilltop hospital behind Salonika.

The British used so much barbed wire to protect themselves that their bastion was called the 'Birdcage'; a small area crowded with refugees of many nationalities as well as French and British troops. The setting was picturesque. Emslie—nicknamed 'Jock'—and her col-leagues could look down at the bay, where a dozen or so hospital ships were anchored, painted white with a red cross and a band of green. They looked particularly attractive at night, when their coloured lights were reflected in the rippling water. But while the Canadian authorities poured money into their nearby hospital, Emslie's team struggled with a wooden hut for an operating theatre, unpaved paths and tents that afforded no protection against either bombs or floods. Throughout the region, women fumed at the parsimony of the Scottish organizing committee:

> I should like to send out here some of the good ladies who sit smugly by their fires and forbade us to take fur coats etc etc and gave us great coats that are almost useless and blankets made of paper and tents that let in the rain. If I hadn't by the grace of God in spite of the directions of the S.W.H. brought out a pair of riding breeches I should have been utterly undone.[18]

Once the freezing winter was over, they had to contend with blinding heat as well as insects carrying disease—ants, flies, mosquitoes. 1916 was the hottest summer for twenty years, but as the temperature soared, mosquito nets and suitable clothing failed to turn up. Water had to be boiled or chlorinated, and food thoroughly cooked to ensure it was safe. But despite all the precautions, the death rate from dysentery and malaria rocketed.

That first year was the worst. As the months dragged on, transport improved, enabling better equipment and drugs to be delivered. French teams began tackling malaria by draining marshes or covering them with petroleum, distributing quinine, and launching a propaganda campaign targeted at the local people and foreign soldiers, many of whom were illiterate. Emslie's patients were housed in wooden sheds provided by the French, while the staff lived in Indian-style tents and had acquired sheepskin coats that doubled up as blankets. She noted that, counter-intuitively, the toughest survivors tended to be thin city-bred women rather than athletic country girls.

Like many women in war-wracked hospitals, Emslie lost interest in what was happening in Britain and paid scant attention to supposedly momentous events such as extending the vote to women. Apart from a brief return home on leave, when she ate enthusiastically with little sympathy for Edinburgh complaints about food rationing, Emslie spent three years in Salonika caring for Serbian soldiers in a Scottish hospital directed by the French Army. Because of the weather, fighting took place only in the spring and autumn, the rest of the year being devoted to medical work. Thanks to suffrage generosity, their laboratory was the best equipped in the region and became a leading research centre. The only one with a microtome (needed to prepare thin samples for microscopic examination), it was also supplied with guinea pigs from Egypt by a friendly Red Cross colonel (they had been a favourite experimental animal since the seventeenth century). To boost morale, in 1917 Hilda Lorimer, a classics don who had taught Vera Brittain at Oxford before volunteering, wrote and staged a Greek play. The following year it was Emslie's turn, and she organized a concert. Every day after work, the cast braved the heat to practise Morris dances, Scottish reels, and Greek-style ballet. Forty Scottish pipers turned up on the night of the event (returning to the trenches the following morning), and Emslie performed a French folk dance in moonbeams simulated by acetylene headlights.

In the summer of 1918, Emslie successfully applied to become the commanding officer of a hospital funded by American donations and based in Ostrovo, ninety miles west of Salonika. 'Just fancy me a C.O. at my tender years,' she wrote proudly to her mother; 'I should have been 20 years older & worn hob-nailed boots & flannel.'[19] Initially, the staff were worried about being supervised by someone so young, but the camp's electrician and driver, Rose West, judged her to be 'a perfect darling...very smart & a great sport'. Although ready to behave like a CO when necessary, West told her mother, 'we always keep smiling—whereas other members are always growling & don't know how to make the best of things' (Figure 14.3).[20]

Located near a beautiful lake, their camp was surrounded by grassy hills studded with wild flowers. Shaded by trees in the summer, the medical staff kept warm during the winter by installing charcoal braziers in their snug green tents. Serbian orderlies carried out the heavy work, which prompted Emslie to wonder whether perhaps the Scottish Hospitals had been a little too insistent on employing only women. Life seemed almost idyllic—apart, that is, from the sound of gunfire and the swarms of mosquitoes that made this a singularly deadly site. Emslie's two predecessors had been invalided home with recurrent malaria, and the staff turnover rate was high.

Emslie decided to move everybody to a safer spot, but she never got the chance. As the Serbian soldiers pushed back into their own country, the Bulgarian forces yielded and her unit was ordered to move deep into reoccupied territory. Exhilarated at the prospect of being nearer the action than any other hospital, the women prepared to leave, ignoring warnings that they might be trapped behind enemy lines. Although given the option of staying back, none of them did.

Figure 14.3. Wounded Serbian soldiers with medical staff at a hospital camp of the 7th Medical Unit of the Scottish Women's Hospitals for Foreign Service, at Ostrovo, Macedonia, during the First World War.

Serbia and Sebastopol

The railway had been destroyed, so Emslie made a 400-mile reconnaissance trip in a little Ford to see what was needed, and on her return headed a convoy of nine vehicles packed with food, medical supplies, and nursing staff. During the last four years, all the women had witnessed appalling devastation and misery, but nothing matched what they encountered now. As their wheels spun in axle-deep mud, they were passed by Bulgarian refugees and bewildered Serbian soldiers plodding along between piles of discarded ammunition. Never again could Emslie see a jay without shuddering to remember the birds pecking at the decaying corpses of donkeys and horses. On the fifth day, as the snow swirled around them, they knew from the stench that they had arrived. Priests mumbled the last rites as they wandered among the hundreds of injured soldiers lying on a stone floor, still in their uniforms, swarming with maggots and lice. Patients wailed continuously as surgeons operated without anaesthetics on a deal trestle table; Emslie never forgot 'the floor swimming in blood . . . the pails crammed with arms and legs and black with flies'.[21]

Sanitation was of paramount importance. It had become a standard joke that whenever the Brits got together in Serbia, their conversation began with lice and ended with latrines. The women immediately installed incinerators, washed the woodwork with paraffin, cleaned up the water system, and began peeling off the men's ancient, blood-soaked bandages. Even after forcing the slightly less sick to leave, they had 450 patients suffering from wounds, dysentery, and the virulent Spanish flu that killed so many healthy young men. The housekeeper reported that Emslie

> looks such a young C.O., but she is most capable, and has made won-
> derful strides to bring order out of a colossal chaos. . . . [W]e had to tackle
> a Herculean task to battle with indescribable filth and vermin, evil smells,
> no rations, no lights, a hospital full of ill and dying men, and everyone
> tired out.[22]

On top of converting an old barracks into a clean hospital, she spent much of her time in bureaucratic nagging to ensure their food supplies. And as well as all that, she found herself responsible for local civilians in villages up to fifty miles away. Most Serbian doctors were either dead from typhus, recuperating in the south of France, or opening lucrative practices in Belgrade.

Constantly busy, the women had little time to ruminate on the horror. Three weeks after they arrived, the Armistice was declared, but they hardly noticed it. For them, the day's exciting news was that Rose West switched on the hospital's new electricity system. Years later, Emslie's main memory of 11 November was the sight of starving Bulgarians fighting over a stolen bucket of chicken scraps. The death rate continued to be high, but burial was difficult in the freezing conditions, and survivors immediately stripped a fresh corpse to clothe the living. The eleven doctors, forty nurses and a handful of administrators struggled on until Christmas, when work continued as normal but—somehow!—turkey and plum pudding were on the menu.

By the time of the Serbian celebrations of Epiphany in January 1919, Emslie had created a well-equipped and spacious hospital, complete with an X-ray department, laboratory, and dispensary, all immaculately maintained by a team of grateful former patients. Every day, around three hundred outpatients arrived. Some of them were local villagers, victims of burns or wolf bites, but others were emaciated Bulgarian prisoners who often died in the waiting room. Their first typhus case turned up in the middle of February, and the epidemic lasted until April, spread by lice that seemed impossible to exterminate. Nurses became sick despite their rubber suits and gloves, although there was one positive result: inside the wards, Serbians, Bulgarians, and Britons abandoned their national differences to care for one another as equals.

Life seemed better in the spring, when the sun banished the typhus and opened the flowers. The railway connection was restored, bringing in visitors and food, and taking away the military men who were no longer needed. After disinfecting the typhus wards, Emslie began accepting civilians needing surgery. Instead of patching up wounded soldiers, her team improved the lives of ordinary Serbians by remedying problems that had afflicted them for many years. Some of them were congenital—cleft palates, webbed hands—but others resulted from accidents or neglect. One young woman had a tumour in her armpit that stretched to the ground and was covered in sores; a little girl's chin was stuck to her chest; a man's fingers grew into his palms. Ninety per cent of the children they saw had tuberculous glands in their neck; Emslie remarked that as none of them drank milk, this scotched the contemporary bovine theory of infection. Later, the women erected tents on the lawn for patients with TB, and as they inherited equipment from units that were closing down, they built a new outpatients' centre in a nearby town.

Figure 14.4. Serbian stamp commemorating Isabel Emslie Hutton, 2014.

For the Serbians, Emslie was more than just a doctor, and—unlike in Britain—she is still honoured there today (see Figure 14.4). Her cures seemed almost miraculous, and they welcomed her into their homes, their lives, and their dancing. She was awarded the Order of the White Eagle, and in 2011, Vranje's new medical school was named after 'Dr Izabel Emsli Haton'.[23] Emslie warmly reciprocated their admiration, learning the language and writing evocatively about their customs. Despite the atrocities, she felt rewarded, and dreamed of establishing a permanent Scottish Women's Hospital. 'I'm operating practically the whole day long—& never had such a wonderful time for work in my life & all on my own too,' she wrote to her mother; 'I can't bear to be under anybody & I'm afraid it'll spoil me for ever afterwards.'[24]

Emslie described her colleagues as a large happy family, but perhaps inevitably, there was friction between the organizing committee of the Scottish Women's Hospitals and the doctors working in the field. Back in Edinburgh, the central Committee decreed it was time for the unit to leave, and on 16 October 1919—after considerable negotiation—she reluctantly obeyed orders and left for Belgrade to take over its Elsie Inglis Memorial Hospital, which Emslie felt was 'a perfect disgrace, and far too small to be of any good.' She wrote home vituperatively that although the Serbians were keen to retain her excellent hospital in Vranje, the committee 'want the Hospital for their darling "Elsie" to be in the Capital where all will see it'. After two Scottish inspectors visited her in Belgrade, she reported contemptuously: 'poor old girls they are such genteel old Edinburgh West Enders that I am very sorry for them—such fearful bores without a single original idea in their heads...they were so inexperienced about everything.'[25]

Emslie also had to face local British resentment. Supporters of her predecessor, Dr Louise McIlroy, who had resigned in disgust at the way she was being treated, maintained that the committee members 'seem incapable of knowing their own minds about anything... & have taken Dr McIlroy's resignation without a murmur & calmly offered the position to Dr Emslie... who is very clever & fascinating—just a flirt— a bacteriologist—no academic standing at all'.[26] The situation deteriorated rapidly. By January, Emslie had discovered that she was being paid less than McIlroy. In February, she fired off a long, angry letter accusing the Edinburgh organizers of ignoring her advice, concluding sarcastically that what they really needed was 'a young Graduate who knows no Serbian and understands nothing of Serbian ways for she will be able at least to satisfy her Home Committee'.[27] Relationships broke down irretrievably in March 1920—and amidst a flurry of angry letters on every side, the hospital closed in April.

Instead of going straight home, Emslie decided to work with children in the Crimea. On her way there, she stopped off for a few days sightseeing, first in Sofia and then in Istanbul (Constantinople for her), where she met an old acquaintance, Major Thomas Hutton, a 'deceptively mild, quiet Englishman'. For once she relaxed, sailing on his yacht and touring the city, but was taken aback one day when he interrupted his intense contemplation of a mosque by proposing to her. She refused, feeling that she was 'more than ever wedded to my profession, which seemed to satisfy me completely'.[28] Continuing her journey, she steamed into Sebastopol on 22 June to take charge of a children's hospital set up by an English aristocrat.[29]

Despite the shortages of money and food, those first summer months were blissful, but Emslie knew that typhus would invade once the weather became colder. She was right to worry: as the snow began falling, the malnourished children became sicker, the military situation worsened and the currency plummeted. On 11 November, exactly two years after the Armistice, she was sent back to Constantinople with her small patients. Thousands of Russian refugees were stranded on ships in the harbour, but after doing everything she could—caring for them medically and dancing to raise money—she decided that she had finally done enough. Setting off for home, she first travelled to Vienna and caught up with the latest medical research.

Isolated on the edge of Russia, Emslie had hardly realized that Britain was slowly returning to normality. After five and a half tumultuous years abroad, she slipped back into her old post at the Edinburgh mental hospital, trying to pick up her former life as if she had never been away.

Married Love

But Emslie *had* been away, and she had changed. Renouncing her long-held vows of a single life, the following year—1921—she married Thomas Hutton, her Army suitor from Istanbul. She was now an expert surgeon who had coped with complicated chronic cases as well as urgent battle injuries, but she faced up to reality: as a woman, there was no chance of practising surgery in Britain. Moreover, although she was able to vote and had run entire hospitals on her own, as a married woman she was obliged to resign from her hospital post and become her husband's financial dependent.

Isabel Emslie Hutton (as she now called herself) embarked on a bleak couple of years living at a military college in Surrey, isolated as a housewife while her husband studied for promotion. After a few months, she managed to escape by setting up a private psychiatric practice near Harley Street. Before long, she had landed a research position to investigate the effects of adrenalin, but further job offers were withdrawn as soon as she revealed that she was married. As she put it, 'the Gates of Paradise . . . soon clashed in my face'; perhaps it was no coincidence that this image featured only a few paragraphs after she had recounted her horror on finding that London hospitals still retained locked doors and padded cells.[30]

Like other professional women who could afford to do so, Hutton practised medicine by volunteering for unpaid positions. Psychiatry was still regarded with suspicion, and there were very few facilities for outpatients: the mentally ill either struggled on or were forcibly confined in overcrowded asylums. At last, in 1925, she saw an advertisement for a post she wanted—honorary consultant psychiatrist at the British Hospital for Mental and Nervous Disorders in Camden. Founded in 1890, this voluntary-run hospital provided treatment for the poor, but mainstream doctors sneered at its shabby buildings and unorthodox methods. As the first woman consultant in a British mixed hospital, Hutton worked there for thirty years, playing a key role in the introduction of psychoanalysis to Britain. She wanted to stay longer, but was squeezed out by new National Health Service regulations.[31]

Never one to flinch from a fight, throughout her post-war career Hutton campaigned to ensure that other married doctors fared better. In 1928, when London medical schools wanted to resume their former status as male-only institutions, one justification was that 50 per cent of qualified female doctors never practised because they got married. Emslie despatched an angry rejoinder to the *Times*, pointing out that the obvious solution was not to ban women but to remove the marriage embargo. Surely, she retorted, 'the wonderful thing is that so many continue to work in spite of the difficulties put in their way.' After all, she went on, perhaps Britain has its own Marie Curie who has been barred from her laboratory? Nancy Astor, the first female MP, headed the list of eminent women who sent a letter endorsing Hutton's protest—might Ray Strachey, who worked with Astor, have been its ghostwriter? Objecting as taxpayers, they demanded 'that in return for our contributions women should be concerned with the management of hospitals, women patients should have the opportunity of being treated by women doctors, and finally that women students should have facilities as adequate as have men students for the purposes of training.'[32]

Hutton was also keen to liberate women at a more personal, intimate level. As a naive medical student, she had been profoundly shocked when a demonstrator suddenly embraced her passionately. Always carefully chaperoned, she knew nothing about sexual relationships, and lacked romantic films or TV programmes to guide her on suitable behaviour. Doctors received minimal training in normal female functions, and they learnt nothing whatsoever about 'moral hygiene'—the euphemism for sex in all its joys and complications. From talking to patients, Hutton realized how desperate women were to know more (or even something) about everyday experiences such as female sexual gratification, birth control, infertility, the menopause, and menstrual abnormalities.

This was, of course, the need that Marie Stopes had tried to meet. But although *Married Love* was a runaway success in terms of sales figures, her readers' letters reveal that its flowery prose often prevented basic information from being correctly understood. In contrast, although Hutton concurred that marital harmony and children's welfare depended on sound sexual relationships, she wrote with clinical detachment and had no qualms about giving clear detailed descriptions. Such frankness, her friends warned her, risked compromising her career. In 1909, even the NUWSS had not dared to publish *Under the Surface*, in which Dr Louisa Martindale described the clinical features of sexually transmitted diseases and argued that prostitution would only cease when women became economically independent.[33] Long after the War, many people still thought it was wrong to discuss such matters openly, and publishers were reluctant to accept Hutton's manuscript for *Hygiene of Marriage*. One popular newspaper refused to advertise it—'We are sorry,' wrote one editor, 'but you must know that we never touch sex stuff.'[34]

Despite defying convention by speaking out, Hutton conformed to prevailing views about gender relationships by giving primary responsibility to the man. In places, *Hygiene of Marriage* (1923) is painful to read, because although she advises like a doctor, her writing is permeated with her own experiences as a new wife. The misery of those first months when she had lost her career and her independence was clearly compounded by sexual anxiety afflicting both partners. They never had children, so perhaps they were both content with what was known as a 'companionate marriage'. On the other hand, she adored children and was firmly opposed to contraception except on medical grounds, so maybe that was a personal sadness. She declared herself delighted to acquire 'my own large family' when she threw herself enthusiastically into running a centre for blind and mentally disabled children.[35]

Over the years, Hutton continued to write similar self-help manuals for women. Although sales were low, she answered many letters from anxious correspondents, and pursued the cause of female equality in every way she could. For example, she was active in the British Women's Medical Federation, which focused on securing equal pay for doctors but also campaigned on wider issues, and in 1926 she volunteered to investigate the ban on women applying for an international pilot's certificate. Aircraft were still so recent that very little medical literature had yet been published, but supposedly, women were considered incapable of flying during periods and pregnancies— a myth she managed to overturn.

'I may say the true success of this stunt is to keep smiling': those words were written by a friend in Vranje to describe Hutton, clearly the sort of woman who just buckled down and made the best of whatever life threw at her.[36] 'We who had taken part in the war were trying to make new lives for ourselves,' she reminisced, 'and some, like myself, had found it heavy going.'[37] After grabbing the opportunity to volunteer for an arduous anti-malarial mission in Albania, in 1938 Hutton dutifully abandoned her own career when her husband was posted to India. Trapped there during the Second World War, she set up her own medical clinics, but—once again—encountered the same official resistance to female doctors as had Inglis in 1914. And once again, just as in Britain a quarter of a century earlier, women began setting up their own voluntary organizations, and Hutton became Director of the Indian Red Cross.

In a private meeting, Mahatma Gandhi told the Hutton husband and wife that hers was 'by far the more important profession, for his is but butcher's work'.[38] Yet it is the military feats of the butchers that are remembered, not the medical research, remedial surgery, and psychological care provided by Isabel Emslie Hutton and her colleagues. Her story is extraordinary, but her activities were not unique: many other female doctors travelled to the Eastern Front. Collectively, they not only restored to health many thousands of soldiers of different nationalities, but also permanently improved conditions in those war-wracked territories.

PART V

CITIZENS OF SCIENCE IN A POST-WAR WORLD

15

INTERWAR NORMALITIES

Scientific Women and Struggles for Equality

[I]t is the achievements of science that, by bringing the ends of the earth together, have raised up new enemies and made possible the slaughter of the human race on a scale undreamed of by our fathers. We are not yet at the end of the achievements of science; and it remains to be seen what measure of personal and political freedom is compatible with our growing knowledge of the secrets and mysteries of nature.

Cicely Hamilton, *Time and Tide*, 12 August 1921

In the British general election of 14 December 1918, more people could vote than ever before. As a major break from the past, men no longer needed to own property: being over twenty-one was the only qualification they required. And for the first time, more than eight million women were deemed to be responsible citizens, although their eligibility was hedged around with restrictions—notably, they had to be over the age of thirty. Because of wartime casualties, there were around eleven women for every ten men, yet they formed well under half the expanded electorate. After all those years of campaigning, political engagement was high. In the next election four years later, turnout was a massive 73 per cent, many of the voters women who turned up immediately the polling booths were opened and—to the amazement of a *Times* journalist—demanded information 'on matters not usually considered women's questions. Foreign policy was a strong point moving women in constituencies.'[1]

Despite such evidence of female commitment, sceptics worried that newly enfranchised citizens were unequal to the task, and they called for better education to narrow the knowledge gaps dividing the genders and the classes. Enhanced by wartime technical developments, public radio was seen as the ideal new medium for teaching electors about their responsibilities: by 1939, 70 per cent of British households owned a wireless set. As head of the BBC, John Reith adopted a top-down attitude, seeing it as his duty 'to give the public what we think they need—and not what they want, but few know what they want, and very few what they need'.[2] But it was a woman, Mary Adams, who decided what the nation would learn about science.

Mary Adams became one of Britain's most influential scientists, promoting science not only as a fascinating subject to study but also as a worthwhile government investment. Recruiting distinguished speakers, she made scientific topics accessible by focusing on innovatory ideas rather than factual knowledge, even enrolling listeners to take part in mass experiments. A BBC colleague enthused that she 'raised high the level of broadcast science talks, through her contacts with scientists at the universities, and her ability to pick out the latest scientific developments and have them presented in a lively and informative way'.[3]

With the help of wartime scholarships, Adams had begun her scientific career as a biology researcher at Cambridge, and in 1928 she gave her first broadcasts on heredity, subsequently published as a book. A keen eugenicist, she emphasized the need to improve the nation's population by encouraging the right sort of mothers to breed. Like Stopes and Emslie Hutton, she advocated providing clearer information (although she could be unconventionally frank—one shocked listener hoped that 'no member of the opposite sex had been obliged to remain in the studio while [Adams] was broadcasting'). In charge of the BBC's education department, Adams organized carefully scripted discussions between eminent scientists such as Julian Huxley and Cyril Burt. Media technology was expanding, and in 1936 Adams became the nation's first female television producer, later responsible for persuading a diffident David Attenborough to try his hand at making wildlife programmes.[4]

Now forgotten, Adams exemplifies those enterprising women who cleverly constructed careers by taking advantage of new technologies that had not yet been fixed as a male preserve. In contrast, conventional scientists, doctors, and engineers discovered that their opportunities closed down once the War was over. Initial gratitude was replaced by popular resentment against women: 'to be frank,' sighed a *Times* journalist in 1919, 'the public has grown tired of uniformed women.'[5] As unemployment rose, priority was given to finding work for men, while women who had successfully performed high-powered jobs were steered back towards domesticity. To modern ears, a Ministry of Labour promotional leaflet sounds like a parody:

> A call comes again to the women of Britain, a call happily not to make shells or fill them so that a ruthless enemy can be destroyed but a call to help renew the homes of England, to sew and to mend, to cook and to clean and to rear babies in health and happiness, who shall in their turn grow into men and women worthy of the Empire.[6]

Opposition also came from employers. Resembling a naïve Darwinian, one senior industrial manager told the *Daily News* that the

> average woman does not possess the same engineering instinct as the average man. For repetition work, yes, but for originality and research, well, there is something lacking—perhaps it is the survival of the Cave days, when the woman stayed at home and the right to live depended upon the man's wits.[7]

Retaliating against such sentiments, the retired business entrepreneur Dorothée Pullinger commented sarcastically that 'I am a member of the Institute of Professional Engineers and also have worked most of the time in Engineering, till I was told I was doing a man out of a job, so I went washing.' Her reference to doing other people's washing—traditionally an unskilled but arduous female job—was a bitter joke. During the War, Pullinger had enjoyed a high-up position in a large munitions factory, and in 1919 she joined the directors' board of an extraordinary women's car manufacturing company that she had persuaded her father to set up in Scotland. Subsequently squeezed out of it, she pragmatically disguised her engineering expertise by wrapping it in a veneer of femininity to run a commercial steam laundry.[8]

Many women who had enjoyed responsible scientific positions during the War were forced into accepting low-status posts. Employers made it clear that 'the question of the ability of women may be left out of the argument; it may be accepted without reserve.' What they really felt was that 'the employment of women is only desirable where the supply of men is deficient.'[9] Very often, less-skilled jobs became redefined as jobs for women, thus reinforcing the notion that only men were capable of climbing to the top of a professional career ladder. As one frustrated graduate complained, 'for women in the chemical industry magnificent health and a thick skin are more important than a knowledge of chemistry.'[10] Even in traditionally female fields, women were generally no better off than before. Around 80 per cent of female Oxbridge graduates became teachers, yet their average wage was around three-quarters of that for men at the same level. In the face of discrimination, low wages and high unemployment, so many female graduates went into teaching that by the 1930s the profession was flooded.

Despite these restrictions, single women could take advantage of their wartime skills to become self-sufficient instead of relying on male protectors. Interviewed in 1925, 'Miss All-Alone' described how she enlisted the help of a female electrical engineer to rewire her little flat, installing light, heat, and a geyser. 'Yes,' she reported nonchalantly (presumably savouring the reaction of the journalist), 'it was a bit of a job running the leads up all those stairs. . . .'[11] For academic high-flyers without a man, the single-sex colleges at Oxford and Cambridge provided an ideal refuge, and some female scientists secured influential positions. For over forty years, the zoologist Honor Fell was funded by the Royal Society to run a Cambridge research laboratory. Although she was sympathetic to the career problems faced by mothers, Fell openly revelled in her personal freedom from family commitments. 'I had a lovely Saturday afternoon, with the whole lab to myself,' she told a scientific friend; 'On Sunday I shall have an orgy of staining slides.' Internationally renowned, Fell's laboratory was organized in a democratic, non-hierarchical way, with a blurred boundary between domestic and scientific settings. Discussions took place while tea was being served, generating a cosy atmosphere with the added bonus of moisture-laden air to prevent tissue cultures from drying out.[12]

Like Ray Strachey, many campaigners continued to insist that achieving economic parity was easily as important as securing suffrage. Whether in the home or the workplace, they argued, women should be paid so that they could lead independent lives. Gaining partial enfranchisement was an important step along the road, but the final destination of equality had not yet been reached. These twentieth-century pioneers fought not to achieve 'great things from counting female noses at general elections, but because the agitation for women's enfranchisement must inevitably shake and weaken the tradition of the "normal woman"'.[13]

Those ironic scare quotes signalled financial and emotional dependence, first on fathers and then on husbands. Normality can mean how most people behave—but more often, it denotes how society deems they *should* behave. Proper girls dressed demurely, learnt how to ensnare a man, and then cared for their children; their greatest intellectual feats involved reading poetry or playing the piano. When a woman became a scientist, doctor, or engineer, she donned the cloak of abnormality by rebelling against expectations that had been reinforced since birth. The battle for the vote had been partially won, but when dealing with daily reality, being a citizen did not necessarily make much difference.

Women and Society

Suffragists had focused on electoral reform, but the first announcement of success in March 1918 was somewhat of an anticlimax. Many women were more preoccupied with winning the War than gaining the vote. 'What use was the vote as a weapon against German guns, submarines and Gothas? The problem of the moment was to keep ourselves alive,' protested a former suffragette stationed in France; 'at this moment of achieved enfranchisement, what really interested me was not the thought of voting at the next election, but the puffs of smoke that the Archies sent after the escaping plane.'[14] In any case, women still had not attained the suffrage goal of electoral parity with men—another ten years went by before the Equal Franchise Act of 1928. Although the government appeared to be recognizing women's ability to participate in running the country, it had effectively disempowered young suffragists keen to retain their wartime independence. Women over thirty were more likely to agree that priority for scarce jobs should be given to men, who earned more money for the same work.

Now that the nation was no longer united against a common enemy, earlier peacetime concerns about class and gender differences reappeared. But however much traditionalists argued for a return to the old conventions, the upheavals of wartime had proved that social change *was* possible. To some extent, class distinctions had been buttressed by the War's military and industrial hierarchies: in 1930, a maid working in a female university hostel was reprimanded for daring to chat with a student.[15] But gender boundaries were more severely challenged, and there could be no going back to exactly the same situation. Although women had always been schooled for marriage and domesticity, the possibility of economic independence threatened to undermine conventional asymmetry.

Feminist organizations were split between two routes to future improvement: should they campaign for protective legislation so that women could feel fulfilled in their time-honoured roles at home, or should they insist on professional parity? According to one camp, women were inherently different from men and should be paid a salary for bringing up children and looking after the home. Faced with the difficulties of getting work in a period of high unemployment, many campaigners supported this notion of wages for domesticity, which garnered broad support for feminist movements by appealing to working-class women. Since the male population had been depleted, and the birth rate was falling, campaigners could give this option a positive spin. 'My own dream and my own vision are of woman as the saviour of the race,' pronounced the Countess of Warwick; 'I see her fruitful womb replenish the wasted ranks, I hear her wise counsels making irresistibly attractive the flower-strewn ways of peace.'[16] This line of argument also strengthened the case for family allowances.

> If women in general, who necessarily include mothers, are to make munitions, serve their country...in any capacity, high or low, it is necessary for our national future that special steps be taken to guard the children who are the natural, sacred and supreme care of womanhood. Whether it be the child of the munition-worker, or of the most highly-placed women-worker in the land, the problem is one and the same.[17]

In contrast, women who had tasted relative freedom during the War, and those who were unmarried, were more likely to believe that they should have the same opportunities and the same status as men, with equality in the workplace. In the eyes of many pre-war suffragists, campaigners for domestic salaries were taking a step backwards. In 1935, the journalist Cicely Hamilton lamented that

> the battle we had thought won is going badly against us—we are retreating where once we advanced; in the eyes of certain modern statesmen women are not personalities—they are reproductive faculty personified. Which means that they are back at secondary existence, counting only as 'normal' as wives and mothers of sons.[18]

For almost four years, employers and ministers had repeatedly praised the women running factories, hospitals, and transport systems. At his silver wedding anniversary of June 1918, the king told a procession of 2,500 uniformed women that

> When the history of our Country's share in the war is written, no chapter will be more remarkable than that relating to the range and extent of women's participation.... Some even have fallen under the fire of the enemy. Of all these we think today with reverent pride.[19]

Only two months after the Armistice, national rhetoric had changed: 'a large part of the female population of the country have had the time of their lives ... swaggering about in every kind of uniform. ... [They must return to being] wives and mothers now the men are coming home.'[20]

Seasoned reformers, often former suffragists, were still active but were less visible than their predecessors. Countless committees and leagues beavered away, successfully improving women's lives by practical steps such as opening up the membership of professional societies, but their quietness enabled a new wave of anti-woman sentiment to spread. There were not enough jobs to go around, and women were in the majority. Formerly seen as 'saviours of the nation', wartime workers were now scorned as 'ruthless self-seekers, depriving men and their dependants of a livelihood'.[21] Fearful and self-protective, men reasserted their former positions of economic and intellectual superiority. When a post as secretary to the Royal Astronomical Society was advertised, a retired soldier told the *Times*: 'I was unsuccessful—at which I did not "kick", until I ascertained that an unmarried women *had been given a man's job!* Surely such an appointment, in these times, is unjustified.'[22]

In the post-War world of unemployment and financial deprivation, men sought to seize back their privileged stronghold, complaining that they were being forced to compete for their own jobs by 'surplus women'. The superintendent of the Woolwich Arsenal was just one of many who objected to this pejorative label, pointing out, 'During the war these same women were one of our greatest assets. There was not a single one too many.'[23] By 1921, despite the protests and despite outnumbering men, fewer women were in paid work than in 1911. Many employers had automatically laid off their wartime women to take back returning soldiers, and most explosives factories simply closed down. Responding to Agnes Conway's questionnaire at the Imperial War Museum, the British Dyestuffs Corporation reported that 'during the war we had a large number of women employed at our factory in Huddersfield, many of whom were engaged on plants for the manufacture of nitric acid for the preparation of explosives.'[24] That past tense suggests little consideration for the fate of former munitionettes forced either to stay at home or to accept a substantial pay cut.

Unlike factory owners, well-meaning committee ladies continued to resist the temptation of hiring women at lower rates. Organizers of a special exhibition showcasing women's war work found it alarmingly expensive to hire mathematicians and artists for creating suitable charts. 'We, of *all* people, don't want to find a woman to do it cheaper, which was suggested to me as a possible economy,' a back-room administrator agonized in a private note to Conway.[25] In contrast, one MP spoke for many when he argued in the *Times* that female applicants did not really need jobs. 'There are still a good many young women who, but for the war, would have stayed at home, and who are now at work only for "pin-money",' he wrote; 'They should be replaced by the ex-soldier in all cases where he could do the work, and I believe that to be the case in many places.'[26] Appealing to popular sentiment, he derided women for frittering away their earnings on clothes and make-up, ignoring the fact that many men regarded beer and tobacco as basic necessities. Although he was right that husbands and fathers needed their wages, he failed to recognize that many wives were obliged to supplement the family income by carrying out paid work on top of their domestic chores.

Women may not have gained as much as suffragists had hoped, but Britain could never revert to pre-war conditions: attitudes, expectations, and emotions had altered. When Lady Astor won her seat in Parliament with an unexpectedly large majority, she identified 'a new spirit both in public and in private life which is struggling to get through. By this I mean the spirit of citizenship and of service which was brought out by the war.'[27] Expressed more colloquially, many people—men as well as women—shared the feelings of Helena Gleichen, the adventurous aristocrat who had driven her radiology equipment through Italy's military zones. 'And after the War? What then?' she wrote many years later; 'As with many other people, we were incapable of sitting still, or of resuming a normal existence.'[28]

Although many women lost their jobs, some managed to remain in positions they would never have held otherwise. For example, British Westinghouse had invested in training their wartime employees to test electrical equipment and calculate graphs for designing machinery—and four years later, these skilled and experienced women were still there.[29] Similarly, wartime improvements in factory safety, mechanization, education, and maternal and child health remained in place, even if they were envisaged as investment in the country's future rather than as altruistic welfare programmes. With this heightened emphasis on family welfare came more openness about sexual intercourse and reproduction. The absence of men for four years had precipitated moral panics about unconventional sexual behaviours (of both men and women), but the aim of preventing both sin and disease did at least stimulate better education. Attitudes towards marriage were slowly changing: whereas formerly its prime purpose had been to conceive and care for children, gradually more emphasis was placed on ensuring sexual enjoyment for both partners.[30]

Many women had tasted independence, freedom, and the pleasure of working for a common cause as well as a salary, but soldiers often came back from the front emotionally depressed, temperamental, and confused. Wounded both psychologically and physically, they felt still more emasculated by knowing how successfully women had taken over their jobs during the War. Queuing forlornly in the hope of work, sending off one unsuccessful application after another, they resented the healthy young women who also wanted to work. Verbal and physical violence escalated, as if men had transferred to the domestic front the military aggression that had been drilled into them during the previous four years. Even some women recommended that for the sake of domestic peace, men should be allowed to pick up their jobs and their privileges. Go back to being normal (that word again), Dr Arabella Kenealy advised women; 'Feminist doctrine and practice menace these most excellent previsions and provisions of Nature by thrusting personal rivalries, economic competition and general conflict of interest between the sexes.'[31]

Women and Science

The Armistice marked the end of the War, but winning the female vote was not the end of the battle for equality. As scientists came back home, they resumed their former positions, and women were once more squeezed out of the top jobs. But there had been three major changes. Most obviously, the very fact that women had been so successful during the War transformed perceptions of their abilities and their social roles. Even people who maintained that women belonged at home with the children could no longer justify their arguments by claiming innate female incompetence. In addition, the statistics had altered. Because of wartime training and expanded education, a higher proportion of women now had professional qualifications in science, engineering, and medicine—and there were also more single women demanding to earn their own living. Just as significantly, academic and industrial research programmes were growing as government, military, and private organizations invested heavily in Britain's scientific development.

Opportunities had opened up, but knowing how to behave as a female scientist remained tricky. In 1925, an apprehensive electrical technologist solicited advice from a friend about a professional conference. 'Please tell me, am I a lady or an engineer—what are you? Do you notice that at each of the papers there is a special side stunt for ladies,' she asked; 'I am entirely in your hands to be organised as you think best—either to join in discussion of the papers—to talk ... to the wives—or merely charm in silence. What ho! I've made a new hat for that—don't be angry.' Dashing off hasty notes that travelled back and forth between them like emails, the two women concluded that they should definitely behave not as wives but as engineers—although the most vexing question to be resolved, the one they expressed most anxiety about, was the design of the hat.[32]

Their informal, bantering exchange stemmed from deeper concerns about male and female identities. Aware of the cynics who mocked women's uniforms, these scientific women focused on outward appearance as a symbol of the inner self. From the boardroom to the bedroom, establishing normality became of paramount importance. By one definition, professional women were abnormal because they were rare—but how easy to slip into assuming that they must also be some sort of freak. One disillusioned scientist remarked bitterly that the only women who succeeded in science were 'enforced celibates, predestined spinsters, and women cunning enough to maintain complete secrecy in their sexual relations'.[33] Negotiating such conflicts imposed great emotional pressures on women, who felt pushed to be domestic goddesses as well as superwomen scientists. In 1927, Dr Octavia Wilberforce knew she was targeting a substantial gap in the medical market when she set up a home for stressed women.[34]

Scientific medicine was redefining sexuality, partly because the War had challenged traditional assumptions. By late 1916 up to 40 per cent of fighting casualties were suffering from symptoms that would have been called hysterical if they had been displayed by women. At first, doctors tried to blame the high numbers on poor heredity, but they were stymied by discovering that officers were more than twice as likely as men of the ranks to break down on the battlefield. Those statistics clashed with assumptions that the upper classes were intrinsically of better breeding stock. A new diagnosis was invented—'shell shock'; like Havelock Ellis's recent invention of the word 'homosexuality', it clouded the formerly uncrossable boundary between men and women.[35]

Amid the post-War flurry of allegations and rumours about female promiscuity, homosexual relationships, and the decline of the human race, researchers tried to provide scientific justifications for diagnosing abnormality and prescribing (so-called) cures. More emphasis was placed on the role of chemical hormones in controlling the body, and after Mendelian ideas became accepted, genetic inheritance could be blamed for what was known as deviance or perversion.[36]

In the emotionally charged atmosphere, pre-existing sciences of sexology and psychology flourished more strongly, while Sigmund Freud's collected works appeared in English in 1922.

Yet although scientific theories and economic pressures were combining to shepherd women back towards domesticity, possibilities were opening up for female scientists because of the imbalance in the population and the national expansion of science-based activities. Government and industry poured money into research: science was here to stay not only in Britain's aircraft and armaments industries but also in the private sector.[37] Just as Mary Adams was able to pursue a successful career in radio, the development of which had been accelerated by the War, so too other women found positions in burgeoning enterprises such as pharmaceuticals, fertilizers, artificial fabrics, and paints.

At first, female chemists could dare to be cautiously optimistic about their future. Women at Sheffield who had been trained to a high level in one specific skill were able to continue their routine work after the War:

> That women have been an undoubted success in this branch of industry, is proved by the fact that notwithstanding so many of the men (who are now demobilised) have resumed duty, a large proportion of the women who desired to stay on have retained their positions to the present time.[38]

An enthusiastic book of 1920 urged young women to consider entering the chemical industry, promising them a stimulating, patriotic career that would ensure Britain's continued supremacy over the Germans. Whereas before the War women 'learnt science to teach women what to teach other women who in their turn taught others', that had reportedly now changed. Packed with practical advice about suitable training courses for the mining, dairy, drugs, and glass industries, this guide came complete with admonitions to look clean and tidy at all times.[39]

The greatest opportunities occurred in those rapidly expanding industrial fields that were not automatically seen as male preserves.

In food factories, some women employed on jam and margarine production lines were able to move up into laboratory positions. But even among scientists with high qualifications, all the old stereotypes reappeared: recruitment literature welcomed female graduates not for their brains, but for their 'manual dexterity, their delicacy of touch, their conscientiousness and their willingness to bear with a routine under which most men become impatient'. Rather than being replacements for men as in the War, women were now being segregated into separate career tracks: some advertisements even specified that this open-ended research post was suitable for a man whereas that routine analytical task would satisfy a woman.[40]

For a woman, being a good scientist was not good enough. A careers adviser spelt it out: 'When it comes to a permanent post, to obtain equal chances with a male rival, the woman must be obviously a little better.'[41] National unemployment was high, and the most attractive posts went to men. Female scientists were paid less for doing the same work, and often found themselves stuck in dead-end positions carrying out repetitive tests. After struggling for years in a pharmaceutical company, a disillusioned chemist protested that

> the male graduate … is paid a reasonable salary and, however young, if his university qualifications are good, he is usually given quite a dignified position from the beginning. The girl who worked side by side with him at the university is hard up and constantly humiliated.…She will be happier if she is not too enterprising because then her sense of frustration will be less.[42]

By 1927, women were being openly warned that their industrial prospects were limited: 'the higher positions call for experience in dealing with workmen, and moreover employers realize that, after a year or two, when the experience which a woman has gained in her work is becoming most valuable, her professional career may be terminated by marriage.'[43]

The situation in hospitals was similar. Many medical schools once again closed their doors to women, and the traditional masculine culture reasserted itself.[44] As a female doctor, Isabel Emslie had entered the War with limited surgical experience, but isolated in Serbia and Salonika she had successfully carried out complex operations on wounded soldiers as well as relieving civilians of crippling burns, congenital deformities, and neglected tumours. Despite acquiring great expertise under atrocious conditions, she 'knew that it would have been unwise and unprofitable to make surgery my life's work at Home'.[45] Reluctantly, she abandoned any prospect of permanently entering this male-dominated field.

Former soldiers returned to university expecting to resume their old lives as if time had stood still in their absence. Everybody found it hard to adapt to the new situation. 'We went to our usual lecture,' reported a female student, 'and found the hall full of men, seated, and women standing or sitting on the floor.' Still more taken aback, the male professor postponed his lecture until the women had been removed.[46] As if women had contributed nothing during the War, a Royal Commission recommended that government investment in university research be increased to reward the efforts of our 'scientific men'. This state funding rescued Cambridge's financial situation, but the award bristled with conditions: the number of female graduates was to be kept under tight control, and women were banned from holding key executive positions. Worried about being overwhelmed by women, Oxford followed suit.[47]

Lack of confidence is often cited as a major hurdle young women need to overcome if they are to pursue the same scientific careers as men. But is this just an excuse for exonerating men, for exempting them from the need to change by placing responsibility on the excluded? An Edinburgh undergraduate felt himself 'moved to bitter speech. To say that women have spoiled our University in no way overstates the case.' Similarly, in an official report at the beginning of the Second World War, an academic also blamed women. In a sadly familiar move, he explained that the low female employment rate in universities was women's own fault—they should have convinced thoughtless male colleagues of their worth. According to him, the remedy also lay in their hands: they should either escape by marrying or begin protesting.[48]

16

LESSONS OF SCIENCE

Learning from the Past to Improve the Future

The past is not dead, but is living in us, and will be alive in the future which we are helping to make.

William Morris, Introduction to *Medieval Lore*
by Robert Steele, 1893

To celebrate its 350th anniversary, in 2010 the Royal Society invited a panel of female scientists and historians to choose the ten women who have had the most influence on science in Britain.[1] The emails soon started flying. Being pernickety academics (and here I mean myself), they naturally began by questioning the question: what does 'most influence' mean? One easy choice was Dorothy Hodgkin, pioneering chemist and still the only British woman to have won a scientific Nobel Prize. But should the list include the seventeenth-century aristocrat Margaret Cavendish? In her favour, she was the first woman to enter the meeting rooms of the Royal Society—but, as opponents pointed out, she was often ridiculed (by women as well as men) and produced no long-lasting theories. What about Elizabeth Garrett Anderson? Although strictly speaking not a scientist, she was one of the earliest qualified doctors and a leading campaigner for female education who affected countless lives. Then there was the unschooled but enterprising Mary Anning. Braving the cliffs of Lyme Regis, she collected fossils and also learned enough about palaeontology to be sure which specimens were valuable. But how important was she to science in general? And how did the panel feel about Jane Goodall, whose empathy with chimpanzees made her a media darling—she certainly aroused public interest in zoology, but did she perpetuate stereotypes of women as touchy-feely earth mothers rather than hard-headed scientists?

The final selection ranged from the eighteenth-century astronomer Caroline Herschel to the modern geneticist Anne McLaren, who died in 2007. All exceptional, all determined, all high achievers, these scientific women illustrate remarkable changes in attitudes and opportunities over the last couple of hundred years. Whereas Herschel was the first woman in this country to receive a salary for her scientific research, McLaren enjoyed a distinguished career at Cambridge and the Royal Society.

In the early twentieth century, female scientists were occasionally on university payrolls, but discrimination still prevailed. Marie Curie's close friend, the physicist Hertha Ayrton, won a Royal Society medal for her groundbreaking research into electric lighting, but in 1902 was turned down for a fellowship on the grounds that she was married. At last, in 1945, two women were elected as fellows of London's Royal Society—the crystallographer Kathleen Lonsdale and the biochemist Marjory Stephenson. The official line at the time was that great progress had been made in opening science to everyone. Thus when the wartime poison gas expert Frances Micklethwait died five years later, her obituarist remarked that 'in these enlightened days' it was far more common for a woman to embark on a career in science than in 1898, when she had enrolled as a student.[2] Contradicting that complacent view, during the 1960s Jocelyn Bell, the PhD student who discovered pulsars, self-protectively slipped off her engagement ring every morning before entering Cambridge's Cavendish Laboratory because she was convinced that a fiancée would not be taken seriously. There was, and there still is, a long way to go.

The chemist Kathleen Culhane, whose life spanned the twentieth century, provides a disturbing example to illustrate the problems faced by female scientists, doctors, and engineers. Born in 1900, she went to university in 1918—the year of the Armistice and the female vote— and lived until 1993. Too young to benefit from the wartime boom in scientific opportunities, after graduating she found that the only way she could even get an interview for a research position was by signing herself 'K. Culhane' to disguise her female identity. When that ruse failed, she resorted to teaching, obtaining laboratory experience by working for no pay in her free time. Eventually, she gained an industrial position—but as her employers blithely informed her, at a salary so low that no man would accept it.

Wages were not Culhane's only problem. Employed by a large pharmaceutical company, she was given routine tasks and banned from the staff lunchroom. When some of her experimental results proved different from those of three male colleagues, she was automatically disbelieved—and yes, hers did indeed prove to be the right ones. When she gave a public information lecture about the importance of including vitamins in margarine, a journalist focused on her appearance—a 'pretty girl with blue eyes and bobbed hair'. Astonished to discover that she needed special permission to continue working after she was married, she insisted on remaining until she had children. Imagine her gall when she discovered that her male successor started at a higher salary than her own final one. Culhane's experiences during the Second World War suggest that insufficient lessons had been learned from the previous experience. Despite the labour shortage, it was only after much badgering that she was allowed to become an assistant wages clerk. Even when she became a fellow of the Royal Statistical Society, she earned far less than her male equivalents—and while they travelled first class on official business, she was left isolated in a third-class carriage.[3]

Although no modern female scientist would encounter such a catalogue of put-downs, aspects of Culhane's story remain familiar in the present century. Speakers at conferences—especially session chairs—are predominantly male. Many scientific departments are overwhelmingly male, with exceptions often due to the initiative of a few individuals. Presenters of television programmes on science are mostly distinguished older men—unless, that is, they happen to be glamorous young women. Researchers who take maternity leave complain of being marginalized, and then face the expenses of full-time childcare. Sceptical scientific mothers feel that there is only one way to reach the top: marry a house-husband.

When Culhane was a baby, the electrical engineer Hertha Ayrton declared that she did 'not agree with sex being brought into science at all. The idea of "woman and science" is completely irrelevant. Either a woman is a good scientist, or she is not.'[4] Well over a century has gone by, but are her words any truer now than they were then? The overt differentiation experienced by Ayrton, Culhane, and many other women is no longer legal, but it appears that discrimination continues to be practised. In principle, equality of opportunity is now firmly entrenched, yet the problem of unequal numbers remains unresolved, especially at higher levels. Even a cursory glance at the statistics reveals that although far more young women are reading science subjects at university than ever before, they are dropping out along the route to the top.

When statistics about women in scientific careers are compiled, I count as a failure. Since leaving an Oxford women-only college with a good degree in physics, I have neither carried out research in a laboratory nor been employed as a scientist: although I could have chosen postgraduate work, I opted for a different career route. I decided to include this autobiographical detail in order to make clear my own position in debates that can become vitriolic and laden with personal accusations. For my first historical projects, I strongly resisted invitations to research into gender. Determined to avoid being branded as an ardent feminist incapable of handling the masculine hard stuff, I picked topics such as magnetism and electricity, Isaac Newton and Joseph Banks. Gradually, I came to realize that for me personally, the main reason for studying the past is to understand the present—and the whole point of doing that is to improve the future. And that is why I have written this book.

As Sherlock Holmes was fond of reminding Dr Watson, it is a great mistake to start constructing theories before you have collected all the available evidence. Gender imbalance prevails at every level of science, but the solution to this problem—and whether it even counts as a problem—is hotly contested. We need fewer impassioned arguments, and more informed debates. Some facts are sufficiently incontrovertible to satisfy even Holmes. Indisputably, women used to be denied the possibility of pursuing intellectually challenging careers governed by the same ground rules as men. Under modern gender legislation, such blatant discrimination would be impossible. Yet there are still unequal numbers of male and female scientists, especially at higher levels. Equality of opportunity is legally vouchsafed, but—to reiterate two common but unsatisfactory metaphors—glass ceilings and leaky pipelines continue to present tough challenges for ambitious scientific women.

It seems crucial to understand *why* far more men than women are working at the top levels of science. Simply blaming men, accusing them of bolstering their own position, is too easy an answer. Not only men but also women need to scrutinize their consciences and explore the behaviour patterns and prejudices that they have inherited from the past, albeit unknowingly. Looking back can feel reassuring, because there are clearly dramatic differences between the status of women now and that of a couple of hundred years ago. Even so, continuities remain. These are particularly significant when they are concealed, lying dormant, unrecognized, and therefore unconfronted.

Could it be possible that some residual belief in male superiority lurks deep inside even the most egalitarian of psyches? Caroline Herschel heads the Royal Society's list of the top ten scientific women, yet her self-abasement sounds appalling. 'I did nothing for my Brother but what a well trained puppy Dog would have done.'[5] Unfortunately, however tempting it is to gasp with horror at such a declaration, it seems that even high-achieving women have internalized that feeling of innate inferiority. Blind trials show that, just like their male colleagues, women rank anonymized job applications higher when they believe the candidates are men.

And then there's parenthood. Extremely few British fathers take up the opportunity of shared post-natal leave. But could that be a mutual problem rather than one restricted to men? In 1826, the official journal of the Royal Society, the *Philosophical Transactions*, published its first full academic article by a woman—Mary Somerville's account of her investigations into the magnetic properties of sunlight.[6] Unlike other researchers, she had not been allowed to present it to the Society's fellows herself. Her husband, a doctor who knew little about her work but did possess the necessary qualification of being male, had read her paper on her behalf. The Society paid tribute to this female scholar by placing her marble bust in the foyer, but continued to ban real-life women from entering. Despite being a gifted mathematician, despite writing numerous books and articles, despite being hailed as the 'Queen of the Sciences', Somerville was embarrassed by her own audacity. 'I hid my papers as soon as the bell announced a visitor,' she confessed, 'lest anyone discover my secret.'[7] Today, female scientists are rightly proud of their success—but how many harbour vestigial traces of Somerville's guilty feelings about neglecting her family?

Blatant mockery and explicit segregation may now have disappeared, but concealed prejudice can be both harder to fight and more keenly felt. Today's scientists point to subtle ways in which women are made to feel like outsiders: the absence of female portraits on corridor walls, the paucity of women's works on student reading lists, the near non-existence of senior women delivering keynote addresses at scientific conferences. If a woman fails to get a lectureship or a research position, is it because a male candidate was better or because she was a woman? Although it is easy to become over-suspicious—sometimes the woman is simply not the most suitable applicant—uncertainty often hovers. Before the First World War, suffragists could see what they were fighting against, but modern discrimination is elusive, insidious, and stubbornly hard to eradicate.

ENDNOTES

Chapter 1

1. Tickner, Plate XIIb, Philip Bagenal to Violet Markham, who opposed the suffrage movement (unsourced).
2. LSE: 7/BSH/2/2/3, letter of 5 Dec. 1907.
3. LSE: 7/BSH/2/2/3, letter of 24 Mar. 1908.
4. Both photographs are in LSE: 7/BSH/6/1.
5. LSE: 7/BSH/2/2/3, letter of 24 Mar. 1908.
6. Dyhouse, *No Distinction of Sex*, pp. 11–55. Noakes, *Women in the British Army*, pp. 23–4. Adam, pp. 24–6.
7. Quoted in Dyhouse, 'Driving Ambitions', p. 325. Vicinus, pp. 121–210.
8. Strachey, *Quaker Grandmother*, p. 80.
9. LSE: 7/BSH/2/2/6, letter from Ellie Rendel of 20 Oct. 1904.
10. Woolf, *Diary*, pp. 200–1 (27 Oct. 1928). See Nicholson.
11. Noakes, '"Playing at Being Soldiers"'.
12. Quoted in Dyhouse, *Girl Trouble*, p. 68.
13. *Punch* (1884), quoted in Gould, p. 132.
14. Quoted in Grattan-Guinness, pp. 118–19.
15. Phillips, Ann, pp. 33–4.
16. Tullberg, pp. 155–79.
17. NCC-WW. I am extremely grateful to Newnham's archivist Anne Thompson for alerting me to this manuscript and for compiling a very helpful list of names, dates, and subjects.
18. Millicent Fawcett (in 1919), quoted Thom, *Nice Girls*, p. 1.
19. Brittain, pp. 467–8. For women's war autobiographies, see Tylee, pp. 184–223.

Chapter 2

1. White, p. 28.
2. Kate Parry Frye's diary, 17 June 1913: Crawford, *Campaigning for the Vote*, p. 157.
3. Quoted in Zimmeck, p. 195.
4. Messenger, pp. 26–7; discussed in Pursell, p. 94.
5. Tickner, p. 184.
6. Quoted in Braybon and Sommerfield, p. 24.

7. Imperial War Museum display, 2014. Bostridge, pp. 359–62. http://www. historylearningsite.co.uk/women_in_1900.htm, accessed 22 May 2015.
8. Bostridge, pp. 1–110.
9. Quoted in Adam, p. 42.
10. 'Germany's Aims and Ambition', *Nature*, 8 Oct. 1914, pp. 137–8.
11. Quoted in Downing, p. 277.
12. Quoted in Woollacott, p. 37 (*Queen*, 10 Feb. 1917).
13. McLaren, Barbara, p. 52.
14. IWM EMP.21: letter from Reading Chamber of Commerce, 22 Feb. 1919.
15. Quoted in Braybon and Summerfield, p. 39.
16. Quoted in Braybon and Summerfield, p. 62.
17. Quoted in Braybon and Summerfield, p. 66.
18. Naomi Loughlan, quoted in Woollacott, p. 42.
19. Alec-Tweedie, pp. 41–3.
20. Alec-Tweedie, pp. 41–4.
21. Quoted in Adam, p. 83. She was the first to take her seat, although the first woman elected was Constance Markievic (in 1918, for Sinn Fein).
22. Quoted in Noakes, *Women in the British Army*, p. 49 (from *Common Cause*, reporting on a London meeting of 1914).
23. Noakes *Women in the British Army*, p. 48.
24. Elsie Phare (later the Shakespearean scholar Elsie Duncan-Jones) quoted in Lee, p. 565. Woolf first gave the lecture at Newnham on 20 October 1928, and then at Girton on 26 October. Woolf, *Diary*, p. 201 (27 Oct. 1928).
25. Woolf, *Room of One's Own*, p. 84; Hall, 'Chloe, Olivia', p. 192.
26. Gwynne-Vaughan, p. 15.
27. Wells, *Ann Veronica*, p. 274.
28. Richmond, 'Balfour Biological Laboratory'.
29. *Manchester University Magazine* 51 (1913), pp. 173–5 (quotation p. 174).
30. Gibert.
31. Lee, pp. 341–61; Ouditt, pp. 169–216.
32. Nina Boyle (1915), quoted in Kent, *Making Peace*, p. 35.
33. Alec-Tweedie, p. 2.
34. Brittain, p. 195.
35. Quoted in Braybon and Summerfield, p. 68.
36. Quoted in Ouditt, p. 30.
37. Quoted in Noakes, *Women in the British Army*, p. 59 (*The Observer*, 13 Aug. 1916).
38. Quoted in Noakes, *Women in the British Army*, p. 59.
39. Quoted in Kent, *Making Peace*, p. 82.
40. Philip Gibbs in quoted Kent, *Making Peace*, p. 44.
41. IWM EMP.21: letter from Reading Chamber of Commerce, 22 Feb. 1919.
42. Braybon, 'Winners or Losers'. Thom, 'Making Spectaculars'.
43. Quoted Leneman, *Service of Life*, p. xi.

44. http://womanandhersphere.com/2015/01/06/suffrage-storiessuffrage-walks-the-suffragette-fellowship-memorial-westminster/ (by Elizabeth Crawford, accessed 4 May 2015). Whitelaw, pp. 50–2.
45. Adie, p. 310.

Chapter 3

1. Kenney quoted at http://womanandhersphere.com/category/suffrage-stories/ ('Shooting Suffrage Films That Suffrage Activists Would Have Seen') by Elizabeth Crawford, accessed 23 Oct. 2015.
2. Liddington: How-Martin and Michael Sadler, quoted on p. 107 (from the *Times*).
3. My major sources for this section are Tickner, pp. 182–205; Richards; Bland, pp. 48–91.
4. Letter to Asa Gray of 3 Apr. 1860 at http://www.darwinproject.ac.uk/entry-2743, accessed 31 Jul. 2014.
5. *Descent of Man* quoted in Richards, p. 119.
6. Wells, *Modern Utopia*, p. 138.
7. Quoted in Bland, p. 74 (from *Man and Woman*).
8. Wells, *Modern Utopia*, p. 138.
9. Henry Armstrong quoted in Jones, Claire, 'Laboratory', p. 186.
10. MacLeod and Moseley, quotation p. 323.
11. Oxford and Asquith, p. 83 (Elizabeth Sloan Chesser).
12. Hamilton, Cicely, pp. 41–50 (quotation p. 47).
13. Cicely Hamilton quoted in Whitelaw, p. 120 (*The Vote*, 14 Jan. 1911).
14. Tickner, pp. 182–205 (quotations pp. 182, 189); Richards; Kent, *Sex and Suffrage*, pp. 24–59.
15. Patrick Geddes (co-author Arthur Thomson) quoted in Bland, p. 78 (1889).
16. Quoted in Kent, *Making Peace*, p. 28 (*The Times History of the Word War*, 1915).
17. Quoted (1916) in Kent, *Sex and Suffrage*, p. 45.
18. Connell and Hunt, pp. 25–32.
19. Quoted in Kent, *Sex and Suffrage*, p. 177 (from *Feminism and Sex-Extinction*, 1920).
20. Quoted in Vicinus, p. 262 (from *Sex Antagonism*, 1913).
21. Quoted Adie, p. 25.
22. Mann, p. 70. *The Times*, 28 Mar. 1912 (reproduced in Wright, pp. 165–88, quotations pp. 168–9).
23. Pick, pp. 189–221 (Huxley quoted on p. 221, attacking the Salvation Army).
24. Arthur Foley Winnington-Ingram, quoted in White, p. 67.
25. Eric Campbell Geddes, quoted in Geddes, *Forging of a Family*, p. 255 (from *The Scotsman*).
26. Rayner-Canham and Rayner-Canham, *Chemistry Was Their Life*, p. 36.
27. Quoted in Rayner-Canham and Rayner-Canham, *Chemistry Was Their Life*, p. 36.

28. Tullberg, p. 85 (letter of 1890). See Chadwick for horses and women; this attitude was later transferred to cars (Doran and McCarthy, pp. 31–2).
29. Quoted in Braybon, *Women Workers*, p. 104 (Eleanor Rathbone, *The Times*, 24 Jun. 1916).
30. Robert Lawson Tait, quoted in Rayner-Canham and Rayner-Canham, *Chemistry Was Their Life*, p. 36.
31. Wells, *Modern Utopia*, pp. 121–45 (quotation p. 140).
32. Richardson (quotation from *Review of Reviews*).
33. Rose, June.
34. Quoted in Kent, *Sex and Suffrage*, p. 54.

Chapter 4

1. Liddington, pp. 26–31, 47.
2. http://womanandhersphere.com/category/suffrage-stories/ ('Shooting Suffrage Films That Suffrage Activists Would Have Seen') by Elizabeth Crawford, accessed 23 Oct. 2015. Thom, 'Making Spectaculars'.
3. Tickner, pp. 55–69.
4. http://womanandhersphere.com/2014/11/21/suffrage-storieswomen-artists-emily-jane-harding-andrews/ (by Elizabeth Crawford), accessed 25 May 2015.
5. Relevant literature includes: Caine, *English Feminism*, pp. 131–72; Pugh, pp. 1–31; Sutherland; Tickner, pp. 182–205.
6. Rheta Childe Dorr, quoted in Tickner, p. 184.
7. Relevant literature includes Sutherland, Holton, and Dyhouse, *Making Peace*.
8. Noakes, *Women in the British Army*, pp. 20–38 (Robert and Agnes Baden-Powell quoted on p. 26). See also Robert.
9. Dyhouse, *Girl Trouble*, pp. 70–104 (quotation p. 74).
10. de Havilland, p. 193.
11. Quoted in Noakes, *Women in the British Army*, p. 32 from *Women and War* (undated).
12. Adlington, p. 14.
13. Downing, pp. 1–60 (quotation p. 14). Edgerton, *Science, Technology*.
14. Quoted (with clauses in opposite order) in Brittain, p. 41. For women's war autobiographies, see Tylee, pp. 184–223.
15. Oxford and Asquith, p. 289.
16. Wells, *Ann Veronica*, p. 13.
17. Crawford, *Campaigning for the Vote*, pp. 10–11.
18. Wosk, pp. 89–119.
19. Rose, June, pp. 16–20.
20. Strachey, Barbara, p. 150 (letter from Lady Gatacre, 1896).
21. Quoted in Adie, p. 144 (1895).
22. Doran and McCarthy, pp. 30–5.
23. Jaffé, pp. 34–68.

24. James Douglas, *Morning Leader*, quoted at http://womanandhersphere.com/2014/11/26/suffrage-stories-an-army-of-banners-designed-for-the-nuwss-suffrage-procession-13-june-1908/ (by Elizabeth Crawford), accessed 25 May 2015.
25. Perrot, p. 49.
26. Strachey, *The Cause*, p. 305. For the male suffrage league, see http://spartacus-educational.com/Wmen.htm, accessed 28 May 2015.
27. Conway. Vicinus, pp. 247–80.
28. Whitelaw, pp. 52–6.
29. Smedley, *Crusaders*, p. 146.
30. Mathews, pp. 28–9.
31. Quoted in Harrison, Brian, p. 158 (18 Nov. 1910).
32. Strachey, *The Cause*, p. 319.
33. Quoted at http://womanandhersphere.com/2014/12/03/suffrage-storieswomen-artists-caroline-watts-and-the-bugler-girl/ (by Elizabeth Crawford), accessed 27 May 2015.
34. Mary Lowndes to Philippa Strachey, quoted in Ticknell, p. 20 (letter of 1908), emphasis in original.
35. Perrot, pp. 49, 51.
36. Shindler, pp. 197–8.
37. Jones, *Femininity*, pp. 91–3; Pankhurst, Emmeline, pp. 289–90; Emling, pp. 19–20. Mason, 'Hertha Ayrton'.
38. Bostridge, pp. 44–210.
39. Quoted in de Vries, p. 75.
40. Strachey, *The Cause*, p. 338.
41. Gregory.

Chapter 5

1. Oxford and Asquith, pp. 97–116 (quotation p. 113); Messenger.
2. Quoted in Kent, *Gender and Power*, p. 284.
3. For example, Holton, pp. 134–53 and Kent, *Making Peace*, pp. 74–96.
4. Daggett, pp. 233–4.
5. Edgerton, *England and the Aeroplane*.
6. Doran and McCarthy, pp. 10–13.
7. Quoted in Strachey, *Women's Suffrage*, p. 25.
8. Pankhurst, Emmeline, p. 330.
9. Daggett, pp. 79, 83.
10. Nina Boyle (1915), quoted in Kent, *Making Peace*, p. 35.
11. Quoted in Kent, *Making Peace*, p. 75 (from 'Socialism in the Searchlight').
12. Noakes, *Women in the British Army*, pp. 1–19 (Woolf quoted on p. 13 from *Three Guineas*, referring to the Spanish Civil War).
13. Daggett, p. 52.

14. Kate Parry Frye's diary, 20 Apr. 1915, in Crawford, *Campaigning for the Vote*, pp. 207–8.
15. Bostridge, pp. 239–60.
16. Carden-Coyne, pp. 8, 215.
17. Noakes, *Women in the British Army*, p. 52 (quotation from *The Daily Citizen*, Feb. 1915).
18. Kate Parry Frye's diary, 17 Jul. 1915, in Crawford, *Campaigning for the Vote*, pp. 209–10.
19. Churchill College Archives, GBR/0014/HALL/3/4, ff. 9–13; James, pp. 72–5.
20. Thomson-Price, pp. 35–49; Strachey, *The Cause*, pp. 337–66. Correspondence in LSE: Box FL321 2LSW/F/4/4.
21. Woollacott, pp. 17–36. Drake, *Women in the Engineering Trades*. Cocroft. Dillon.
22. White, p. 103.
23. Johnson. For international comparisons, see Braybon, 'Women, War, and Work', and Grayzel.
24. Quoted in Thom, *Nice Girls*, p. 146 (Ministry of Munitions, 1916).
25. Rebecca West quoted in Dillon, p. 102.
26. West (quotations p. 92). Dillon, pp. 103–4.
27. Adie, pp. 142–50 (quotation p. 145). White, pp. 46–68 (especially pp. 61–2).
28. *Women's Liberal Review* of 1915, quoted in Kent, *Gender and Power*, p. 277.
29. Quoted in Kent, *Gender and Power*, p. 284 (15 Aug. 1916).
30. L. K. Yates quoted in Smith, p. 198.
31. Daggett, p. 63.
32. Quoted in Potter, p. 110, from *Britannia's Revue*, 1918.
33. Thom, 'Making Spectaculars'. Condell and Liddiard describe and reproduce 163 photographs.
34. Quoted in Thom, *Nice Girls*, p. 152 (omitted from the published version of a report on women in industry).
35. Clarsen (quotation p. 334).
36. Quoted Hynes, p. 91 (from her diary).
37. Miall.
38. Quoted in White, pp. 102–3.
39. Joan Williams, quoted in Woollacott, pp. 32–3.
40. Isabella Clarke, quoted in Braybon and Summerfield, p. 69.
41. Quoted in Thom, *Nice Girls*, p. 42.
42. Rayner-Canham and Rayner-Canham, *Chemistry was Their Life*, p. 448.
43. http://www.warpoetry.co.uk/owen1.html, accessed 18 June 2015.
44. G. M. West, quoted in Woollacott, p. 35.
45. Quoted in Adie, p. 139 without source.
46. Quoted in Woollacott, p. 81.
47. Quoted in Woollacott, pp. 81–2.
48. Caroline Playne quoted in Kent, *Making Peace*, p. 37.
49. Advertisement reproduced Woollacott, Figure 5. Doran and McCarthy, p. 76.

50. Thom, *Nice Girls*, pp. 122–43 (quotation p. 127 from the *British Medical Journal*, 1917). Woollacott, pp. 59–88.
51. Dr Janet Campbell, quoted in Thom, *Nice Girls*, p. 197 (1919 report on women in industry).
52. Gabrielle West, quoted in Smith, p. 202.
53. Quoted Smith, p. 201.
54. IWM ED. 2. 16/928: letter from Fred Knowles, University of Sheffield metallurgy department, Aug. 1919.
55. http://collections.vam.ac.uk/item/O86983/acetylene-welder-the-great-war-print-nevinson/, accessed 21 Nov. 2014.
56. G. M. West, quoted in Woollacott, p. 35.
57. McLaren, Barbara, p. 54 ('gaigne' is not in the *Oxford English Dictionary*).
58. M. O. Kennedy, *National News*, 7 Oct. 1917, quoted in Noakes, *Women in the British Army*, p. 73.
59. George Wilby, quoted in Smith, pp. 203–4.
60. Thom, *Nice Girls*, pp. 53–77; Drake, *Women in Trade Unions*, pp. 68–110.
61. Quoted in Adam, p. 59.
62. Drake, *Women in the Engineering Trades*, p. 43.
63. Quoted in Adam, p. 59.
64. Doran and McCarthy, pp. 60–7, 74–5.
65. Adlington, p. 22.
66. Braybon and Summerfield, p. 47.
67. Quoted in Braybon, 'Winners or Losers', p. 92 (Arthur Marwick).
68. Woman correspondent quoted from the *Morning Post*, July 1916, in Kent, *Making Peace*, p. 37.
69. Woollacott, p. 132.
70. Woollacott, pp. 113–33.
71. Quoted in Woollacott, p. 123 (from *Woman's Life*, 1916).
72. Quoted in Thom, *Nice Girls*, p. 154.
73. Drake, *Women in Trade Unions*, pp. 82–3 (quotation p. 83). Society quoted letter of Jan. 1, 1917, one of many in LSE: Box FL321 2LSW/F/4/4.
74. Quoted in Grayzel, p. 129 from *The Bulletin*, 3 Sep. 1917.
75. Thom, *Nice Girls*, p. 20 (from a 1914 letter to the *Southport Visitor*).
76. Quoted in Woollacott, p. 130 (from the *Daily Express*, 1917).

Chapter 6

1. Quoted in Harrison, Brian, p. 180 (Helen Darbishire).
2. LSE: 7/BSH/2/2/1, letter to her mother Mary Costelloe of 14 Nov. 1909. My major secondary sources are: Strachey, Barbara; Harrison, Brian, pp. 151–83; Hamilton, Mary, pp. 218–21, 263–73. For women's war autobiographies, see Tylee, pp. 184–23.
3. LSE: 7/BSH/2/2/2, letter to her mother Mary Costelloe, 11 Nov. 1910.

4. LSE: 7/BSH/2/2/2, letter to her mother of 21 Dec. 1910.

5. LSE: 7/BSH/2/2/3, letter to her aunt of 21 Nov. 1910.

6. Quoted in Strachey, Barbara, p. 254 (letter to Vanessa Bell of 6 Apr. 1911).

7. Quoted in Strachey, Barbara, p. 254 (letter to Vanessa Bell of Apr. 1911).

8. Quoted in Harrison, Brian, p. 160 (letter from Woolf of Aug. 1915).

9. Wyllie and McKinley, p. 253.

10. For a collective biography of the Strachey family, see Caine, *Bombay to Bloomsbury*.

11. Frances Elinor Rendel (Ellie) was the daughter of Lytton Strachey's older sister, Elinor. The diary of her American trip is in Harvard Library; she was later Virginia Woolf's doctor.

12. Quoted in Strachey, Barbara, p. 200 (diary entry of 19 Mar. 1904).

13. Strachey, *Marching On*, p. 7.

14. Photographs and leaflet in LSE: Box 14 7/BSH/6/1.

15. Strachey, *Quaker Grandmother*, p. 109.

16. Quoted in Strachey, Barbara, pp. 201–2 (letter of 6 Feb. 1902). Mary Boole was the widow of George Boole, the originator of Boolean algebra.

17. Strachey, *Women's Suffrage*, p. 22.

18. Quoted in Strachey, Barbara, p. 253 (letter to Ottoline Morrell of 25 Mar. 1911).

19. Crawford, *Enterprising Women*, pp. 52–3.

20. LSE: 7/BSH/2/4/1, private note of 21 Nov. 1904.

21. Strachey, *Quaker Grandmother*, p. 80.

22. Quoted in Strachey, Barbara, p. 201 (letter to Mary Costelloe of 17 May 1901).

23. Quoted in Harrison, Brian, p. 164 (letter of 4 Feb. 1917).

24. Quoted in Harrison, Brian, p. 175 (letter to her mother of 10 Feb. 1933).

25. Strachey, *Quaker Grandmother*, p. 118.

26. LSE: 7/BSH/2/2/3, letter of 25 Dec. 1904.

27. LSE: 7/BSH/2/2/3, letter of 8 Jan. 1906.

28. LSE: 7/BSH/2/2/3, letter of 5 Dec. 1907.

29. LSE: 7/BSH/2/2/3, letter of 7 June 1908.

30. MacLeod and Moseley, p. 327.

31. LSE: 7/BSH/2/2/3, letter of 24 Mar. 1908.

32. LSE: 7/BSH/2/2/3, letter of 7 Jul. 1909. Shaw, quotation p. 217.

33. Quoted Strachey, Barbara, p. 238 (letter of 1 Feb. 1907).

34. LSE: 7/BSH/2/2/3, letter of 23 Jan. 1909. There are several American letters in this folder.

35. LSE: 7/BSH/2/2/3, letter of 21 Nov. 1910. Strachey, *Quaker Grandmother*, p. 143.

36. http://womanandhersphere.com/2013/06/25/suffrage-stories-bloomsbury-links-part-2/ (by Elizabeth Crawford), accessed 26 Nov. 2014. Tickner, pp. 16–18. Grace Darling (1815–42) was a lighthouse-keeper's daughter, celebrated for rescuing survivors from a shipwreck of 1838.

37. LSE: 7/BSH/2/2/4, letter of 28 Apr. 1911.

38. Quoted in Harrison, Brian, p. 174 (28 Jul. 1933).
39. Quoted in Whitelaw, p. 74 (from *Unfinished Adventure*); see also pp. 109–14 and Vicinus, pp. 148–62.
40. LSE: 7/BSH/2/2/6, undated letter, probably 1910. There are also many letters in 7/BSH/2/2/7 and 7/BSH/2/2/8.
41. LSE: 7/BSH/2/2/6, letter of 31 May 1911.
42. Quoted in Strachey, Barbara, p. 287 (Ray Strachey, letter to her mother of 20 Jun. 1926).
43. Quoted in Strachey, Barbara, p. 286 (letter to their mother of 2 Apr. 1924).
44. Harrison, p. 168 (letter from Ray to Oliver of 2 Feb. 1917).
45. Quoted in Strachey, Barbara, p. 289 (letter to her mother of 22 Oct. 1922).
46. Quoted in Harrison, Brian, p. 171 (letter to Lady Astor of 19 Jul. 1940).
47. Harrison, Brian, p. 157.
48. Strachey, *Marching On*, p. 91.
49. Quoted in Harrison, Brian, p. 161 (letter of 1919).
50. Hamilton, Mary, p. 265.
51. Quoted in Harrison, Brian, p. 162 (24 June 1913).
52. Quoted in Dillon, p. 101 from the *Annandale Observer*.
53. Quoted in Harrison, Brian, p. 164 (letter to her mother of 1 Apr. 1917).
54. Quoted in Strachey, Barbara, p. 274 (letter to Millicent Fawcett of 18 Feb. 1918, emphasis in original).
55. For Oliver Strachey, see Wyllie and McKinley, especially pp. 53–6, 241–2.
56. Strachey, *Fawcettt*, pp. 273–301.
57. Quoted in Caine, *Bombay to Bloomsbury*, p. 313 (letter of 21 Apr. 1915).
58. Kent, *Making Peace*, pp. 74–96 (Lloyd George quoted p. 86, March 1917).
59. Quoted in Harrison, Brian, p. 164 (emphasis in original).
60. Quoted in Strachey, Barbara, p. 273 (Ray Strachey, letter to Mary Costelloe of 4 Feb. 1917).
61. Strachey, *The Cause*, p. 360.
62. Strachey, *The Cause*, pp. 361–2.
63. Quoted in Strachey, Barbara, p. 274 (Ray Strachey, letter to her mother of 11 Feb. 1917).
64. Hamilton quoted Whitelaw, p. 190 (*English Review*, Feb. 1921).
65. Quoted in Adam, p. 85.
66. http://womanandhersphere.com/2015/03/16/suffrage-stories-christabel-the-ballot-box-and-that-hat/, accessed 27 Jul. 2015.
67. Quoted in Harrison, Brian, p. 167.
68. Quoted in Adam, p. 88.
69. http://researchbriefings.files.parliament.uk/documents/SN06652/SN06652.pdf, accessed 8 Feb. 2016. Nicholson, p. 325.
70. Woolf, *A Room of One's Own*, pp. 38–9.
71. Strachey, *Our Freedom*, pp. 120–1.
72. Hamilton, Brian, p. 276. LSE: 7/BSH/2/2/4, letter of 1 Aug. 1912.

73. Strachey, *Women's Suffrage*, p. 26. Correspondence in LSE: Box FL321 2LSW/F/4/4.
74. LSE: 7/BSH/2/2/5, letter of 24 Sep. 1918. Powell, pp. 281–91.
75. Strachey, *Our Freedom*, p. 127.
76. Quoted in Kent, *Making Peace*, p. 101.
77. Strachey, *The Cause*, p. 371.
78. Strachey, *Our Freedom*, p. 130.
79. Quoted in Harrison, Brian, p. 165 (from 24 Jan. 1918).
80. Strachey, *Our Freedom*, pp. 133–4.
81. Nicholson.
82. Quoted in Harrison, Brian, p. 173 (letter of 11 Aug. 1927).
83. Quoted in Strachey, Barbara, pp. 299–300 (letter of 8 Dec. 1931).
84. Quoted in Harrison, Brian, p. 174 (letter to her mother of 26 Jan. 1937, emphasis in original). For the Civil Service, see Zimmeck.
85. Quoted in Harrison, Brian, p. 176.
86. Quoted in Caine, *Bombay to Bloomsbury*, p. 319 (letter to her sister Pernel of 13 Jul. 1944).
87. Quoted in Harrison, Brian, p. 178 (letter of 18 June 1940).

Chapter 7

1. Goodman, quotation p. 88 (S. Pembrey to John Scott Haldane).
2. Gay, p. 70.
3. Zangwill, p. 27.
4. Wells, *Ann Veronica*, p. 218.
5. Mann, p. 20.
6. Clark. I am very grateful to Sally Shuttleworth for her comments at a University of Kent conference (July 2016) on Ormerod as an early practitioner of Citizen Science.
7. Browne.
8. Quoted from his obituary, Rayner-Canham and Rayner-Canham, *Chemistry Was Their Life*, p. 433.
9. Henry Armstrong quoted in Jones, *Femininity*, p. 91. See also Mason, 'Hertha Ayrton'.
10. Grattan-Guinness, quotations p. 141.
11. Shindler, pp. 196–206.
12. Rayner-Canham and Rayner-Canham, *Chemistry was Their Life*, pp. 421–45 (quotations pp. 424–5, 437).
13. Stephenson, p. 915.
14. Rayner-Canham and Rayner-Canham, *Devotion to Their Science*, pp. 127–60.
15. Hemmungs Wirtén, p. 76. Article by James Johnston, http://trove.nla.gov.au/ndp/del/article/38832343, accessed 4 Aug. 2015 (*Western Mail*, 12 Aug. 1911, p. 39).

16. Mason, 'Women Fellows' jubilee', p. 232.
17. Fawcett. I am grateful to Douglas Palmer for this information.
18. William Hardy quoted in Richmond, 'Balfour Laboratory', p. 443.
19. Jones, 'Laboratory'.
20. Burek, pp. 9–23.
21. M. D. Ball in Phillips, Ann, p. 77.
22. M. D. Ball in Phillips, Ann, p. 76.
23. Richmond, 'Women in the Early History' (quotation p. 73). Marsha Richmond's two marvellous articles are my major source for my discussion of Cambridge women. See also MacLeod and Moseley.
24. Stephenson.
25. Richmond, 'Balfour Laboratory' (quotation p. 427); statistical table in MacLeod and Moseley, p. 327.
26. Student quoted in Burek, p. 21.
27. Eleanor Sidgwick quoted in Richmond, 'Balfour Laboratory', p. 445.
28. Zimmeck (quotation from a 1904 report, p. 192).
29. Rayner-Canham and Rayner-Canham, *Chemistry was Their Life*, pp. 24–9. Vicinus, pp. 121–210.
30. MacLeod and Moseley, p. 330.
31. Pottle.
32. Rayner-Canham and Rayner-Canham, *Devotion to Their Science*, p. 123.
33. Grattan-Guinness, pp. 116, 117.
34. Quoted in Rayner-Canham and Rayner-Canham, *Chemistry was Their Life*, p. 32.
35. Quoted Dyhouse, 'British Federation', p. 475 (from *The Freewoman*, 1911). For King's College, see Oakley.
36. Creese, 'Benson'.
37. Letter of 11 Nov. 1902, quoted in Green, p. 112.
38. Sutherland, pp. 18–40.
39. Quoted in Forster, p. 11 from an unnamed 'able' scientist.
40. Dyhouse, *No Distinction*, pp. 217–21.
41. Quoted in Rayner-Canham and Rayner-Canham, *Chemistry was Their Life*, pp. 396–7.
42. Morrissey. Chick, Hume, and Macfarlane, pp. 14–30.
43. Dyhouse, *No Distinction of Sex?* pp. 143–4.
44. Mason, 'Forty Years' War'. Article by James Johnston, http://trove.nla.gov.au/ndp/del/article/38832343, accessed 4 Aug. 2015 (*Western Mail*, 12 Aug. 1911, p. 39).
45. Rayner-Canham and Rayner-Canham, *Devotion to Their Science*, p. 157.
46. Bowden.
47. Morley, pp. 11–24 (quotation p. 16); Dyhouse, 'British Federation'. Dyhouse, *No Distinction*, pp. 156–61.
48. Gay, pp. 122–5 (fn. 59 and 81).

49. Murray, Margaret, pp. 158–60 (quote p. 160).
50. Jones, *Femininity*, p. 201, quoted from *The Times*, 16 June 1914.
51. Green; Rose, June.
52. Maude, pp. 65, 68–9.
53. Daggett, pp. 210–11. Bindman, Brading, and Tansey. Rayner-Canham and Rayner-Canham, *Chemistry was Their Life*, pp. 53–8.
54. Burek, pp. 23–32 (quotation p. 29).
55. Rayner-Canham and Rayner-Canham, *Chemistry was Their Life*, pp. 62–94.
56. Dyhouse, 'British Federation'.
57. Creese, 'Sargant'. IWM ED. 3/11.

Chapter 8

1. Bostridge, p. 205.
2. Helen Schlegel in *Howard's End*.
3. Hynes, pp. 3–24 (Gray quoted p. 3; Brooke quoted p. 13).
4. Wyndham Lewis, quoted in Edgerton, *Aeroplane*, p. 13. Bostridge, pp. 131–5.
5. Bostridge, pp. 280–95.
6. Edgerton and Pickstone; Edgerton, 'Time, Money, and History'. I am very grateful to David Edgerton for commenting on a draft of this chapter and alerting me to these articles.
7. Lewis Mumford, quoted in Edgerton, *Shock of the Old*, p. 141.
8. MacLeod, Roy, p. 473.
9. *Nature*, 9 Sep. 1915, p. 34. See also Reid, p. 39 and 'Dear Harry' exhibition, Oxford Museum of the History of Science (visited 14 June 2015).
10. MacLeod, Roy, pp. 473–4.
11. Lord Malcolm Davidson, private communication, 27 Mar. 2015: Sir James Mackenzie Davidson conversed in dog Latin with Roentgen in 1887.
12. Stanley, pp. 59–67.
13. Hynes, p. 34 (1915 cartoon reproduced opposite p. 50).
14. Edgerton, *Shock of the Old*, pp. 138–59. Adie, pp. 246–53.
15. My major sources to counter this argument are Edgerton, *Aeroplane*; Edgerton, *Science, Technology*; Downing. For other interpretations, see Hartcup; MacLeod (Roy); Marwick; Rose and Rose; Freemantle. For medical changes, see Harrison, Mark.
16. *Nature*, 29 Oct. 1914, p. 222. Turner. Downing, pp. 5–8.
17. *Nature*, 8 Oct. 1914, p. 138.
18. Quoted Edgerton, *Aeroplane*, p. 4.
19. Gollin.
20. William Osler, quoted in Hull, p. 467; Cardwell.
21. MacLeod, Christine.
22. Arthur Lee attacking Lord Haldane, quoted in Edgerton, *Aeroplane*, p. 6.
23. Quoted in Downing, p. 155.

24. Katzir.
25. Van der Kloot.
26. Downing, pp. 41–74. Edgerton, *Aeroplane*, pp. 1–17.
27. Downing, pp. 101–23; Wyllie and McKinley, pp. 155–69.
28. Rose, Norman.
29. Hull. http://discovery.nationalarchives.gov.uk/details/r/C89, accessed 1 Jul. 2015.
30. Simmons, pp. 57–86.

Chapter 9

1. Wachtler and Burek.
2. Evans, pp. 186–201, pp. 223–35 (quotation p. 233). Conway.
3. IWM ED. 2.9/4 and ED. 2.16/3.
4. Rayner-Canham and Rayner-Canham, 'British Women Chemists', p. 23.
5. Rayner-Canham and Rayner-Canham, *Chemistry was Their Life*, pp. 463–4.
6. Howson.
7. NCC-WW: pp. 72–81. Rayner-Canham and Rayner-Canham, *Chemistry was Their Life*, pp. 188–90.
8. Pursell, pp. 80–1.
9. IWM MUN. 15.3/3, report by Hilda Hudson. Creese, 'Hudson'.
10. Rayner-Canham and Rayner-Canham, *Devotion to Their Science*, pp. 52–5.
11. Rayner-Canham and Rayner-Canham, *Devotion to Their Science*, pp. 156–7.
12. Rayner-Canham and Rayner-Canham, *Chemistry was Their Life*, pp. 453–4.
13. Green, pp. 107–19, 175–6 (letter of 12 Sep. 1916 quoted p. 176).
14. Maude, pp. 76–120 (Sir Jethro Teall quoted p. 76).
15. Drummond (quotation from plate between pp. 250 and 251).
16. IWM EMP. 64/39: letter from British Westinghouse.
17. Baker. I am very grateful to Nina Baker for sending me this article.
18. Pottle. Quotation from McKillop and McKillop, p. 14.
19. Quoted in Rayner-Canham and Rayner-Canham, *Chemistry was Their Life*, p. 452 (1915). MacLeod, pp. 474–5.
20. Quoted in Rayner-Canham and Rayner-Canham, *Chemistry was Their Life*, p. 449, p. 448.
21. Deacon.
22. Jackson and Jones. My main source for Dorothea Bate and the Natural History Museum is the marvellous book by Karalyn Shindler, especially pp. 196–214 (quotation p. 208).
23. Creese, 'Smith'.
24. Izzard, pp. 79–80.
25. IWM MUNN 15.2/5.
26. IWM EMP. 64/18. Chick, Hume, and Mafarlane, p. 125.
27. Ford. NCC-WW, p. 187.

28. Cohen, pp. 50–5.
29. Brittain, p. 145.
30. Guillebaud, pp. 48–9 and *passim*.
31. http://www.ww1.manchester.ac.uk/the-university-and-the-war/impact-of-war/, accessed 11 Nov. 2015.
32. Phillips, Ann, pp. 79–80 and 99–135 (M. G. Woods quoted p. 103). Cohen, p. 51.
33. Rayner-Canham and Rayner-Canham, *Chemistry was Their Life*, pp. 191, 450–6.
34. My major source of information on wartime Imperial College is Gay, pp. 114–46.
35. J. D. Bernal quoted in Gay, p. 114.
36. Rayner-Canham and Rayner-Canham, 'British Women Chemists', p. 23.

Chapter 10

1. Adie, p. 279.
2. CCC. Bruford papers: MISC 20.
3. Rayner-Canham and Rayner-Canham, 'British Women Chemists', p. 21. See also Creese, 'British Women'.
4. Rayner-Canham and Rayner-Canham, *Chemistry was Their Life*, p. 284.
5. Phillips, Ann: M. D. Ball p. 77 and A. I. Richards, p. 133.
6. Margaret Tuke, quoted in Dyhonse, 'British Federation', p. 477.
7. Quoted in Richmond, 'Balfour Laboratory', p. 439.
8. Quoted in Gibert, p. 412.
9. Quoted in Gibert, p. 412.
10. Gay, p. 421.
11. Wells, *Ann Veronica*, p. 160.
12. Smedley, *Crusaders*, pp. 1–15; Creese, 'Maclean'.
13. Quotations from Bowden and Rayner-Canham and Rayner-Canham, *Chemistry was Their Life*, pp. 22–3.
14. Phillips, Ann: M. D. Ball pp. 76–8.
15. Peck, pp. 13–18; Freedman. I am very grateful to Will Peck for permitting me to refer to his excellent but unpublished dissertation, and to Robert Freedman for sending me a copy of his conference presentation on Smedley.
16. *Manchester University Magazine* 51 (1913), 173–5.
17. Hall, Ruth, pp. 51–9; Maude, pp. 55–6.
18. Article by James Johnston, http://trove.nla.gov.au/ndp/del/article/38832343, accessed 4 Aug. 2015 (*Western Mail*, 12 Aug. 1911, p. 39).
19. Hall, 'Chloe, Olivia', p. 197.
20. Brockington; Smedley, *Crusaders*, pp. 23, 65, 144–7.
21. Smedley, *Woman*, quotation pp. 84–5.
22. Dyhouse, 'British Federation' (Dr Merry Smith quoted p. 472); Bowden; Creese, 'Maclean'.
23. My major sources for Whiteley are Barrett and Creese, 'Whiteley'.

24. I am very grateful to Annabel Valentine, Archivist at Royal Holloway, for sending me this information.
25. Rayner-Canham and Rayner-Canham, *Chemistry was Their Life*, pp. 151–7 (Whiteley quoted p. 156).
26. Quoted in Rayner-Canham and Rayner-Canham, *Chemistry was Their Life*, pp. 122–3.
27. Hemmungs Wirtén.
28. IWM EMP. 64/12–14, 30 Oct. 1919.
29. Gay, pp. 124–5. Rayner-Canham and Rayner-Canham, *Chemistry was Their Life*, pp. 454–6 (quotation from 1953, p. 455).
30. Quoted in Gay, p. 125.
31. My account is mainly taken from Mason, 'Forty Years' War'.
32. Rayner-Canham and Rayner-Canham, *Chemistry was Their Life*, pp. 62–84 (quotation p. 74).
33. William Astbury, quoted Rayner-Canham and Rayner-Canham, *Chemistry was Their Life*, p. 86.
34. http://www2.warwick.ac.uk/fac/arts/history/chm/research_teaching/womenbiochemists/biochemicalsociety/, accessed 31 Aug. 2015 (Professor A. E. Chibnall).

Chapter 11

1. http://www.rsc.org/images/mabel_elliott_obit_tcm18-210098.pdf and http://www.bbc.co.uk/news/education-15621443, accessed 24 June 2015. For espionage paranoia, see Bostridge pp. 296–320 and Emslie.
2. Burstall; Gay, p. 125.
3. Davis.
4. Peck, pp. 4–5. I am very grateful to Will Peck for permitting me to refer to his excellent but unpublished dissertation. NCC-WW, p. 187.
5. Adie, pp. 246–53.
6. Mathews, pp. 11–32. Terry, *Women in Khaki*, p. 31.
7. Churchill College archives, Bruford papers: MISC 20; Downing, pp. 126–8; James, pp. 24–42; Wyllie and McKinley, pp. 228–31; Ralph.
8. Churchill College archives, GBR/0014/HALL/3/1, f. 4.
9. Wells, *Ann Veronica*, p. 274.
10. William Clarke quote from Wyllie and McKinley, p. 230.
11. Fraser, p. 62.
12. Birch, Knox, and Mackeson (quotations pp. 48, 45).
13. Wyllie and McKinley, especially pp. 53–6, 241–2, 300–5 (quotation p. 56).
14. Hay, pp. 58–89.
15. Conway.
16. *The Globe*, 19 Sep. 1917, quoted Noakes, '"Playing"', p. 134. See also Robert.
17. White, pp. 187–9. McLaren, Barbara (quotation p. 10, Introduction by Asquith).

18. McLaren, Barbara, pp. v–vi. Lloyd George quoted Terry, *Women*, p. 30 (without source).
19. McLaren, Barbara, pp. 124–8.
20. White, pp. 109–14.
21. Creese, 'Thomas'.
22. Burek, pp. 21–3. Phillips, Ann, p. 99.
23. M. D. Ball in Phillips, Ann, p. 76.
24. Creese, 'Shakespear'.
25. McLaren, Barbara, pp. 117–19. http://collections.museumoflondon.org.uk/online/search/#!/results?terms=gertrude%20shaw&search=OR%3Btype%3Bobject, accessed 18 Aug. 2015.
26. Mason, 'Bidder'.
27. Stevenson; Gay, p. 184.

Chapter 12

1. Quoted in Terry, 'Watson'. I have found no evidence that Mona and Patrick Geddes were connected: he is mentioned in neither Geddes nor Crawford, *Enterprising Women*.
2. Gwynne-Vaughan, pp. 14, 27, 51. See also Robert.
3. Noakes, *Women in the British Army*, pp. 1–81. Terry, *Women in Khaki*, pp. 32–72 (Haig quoted p. 39).
4. http://blogs.nam.ac.uk/exhibitions/online-exhibitions/waacs-war, accessed 18 Aug. 2015.
5. Noakes, '"Playing"'.
6. Noakes, *Women in the British Army*, pp. 81–93 (quotation p. 85, *Sheffield Daily Telegraph*, 12 June 1919).
7. https://womanandhersphere.com/2015/10/16/suffrage-stories-shooting-suffrage-films-that-suffrage-activists-would-have-seen/, accessed 23 Oct. 2015.
8. My major sources for this account are Terry, 'Watson' and Geddes, Auckland (quotation p. 189).
9. Elston; Somerville.
10. Quoted in Whitehead, p. 120 (about St Mary's in 1917).
11. IV, quoted from *British Medical Journal*, 15 Aug. 1936, p. 371.
12. Quoted without source in Terry, *Women in Khaki*, p. 23.
13. Gwynne-Vaughan, p. 13.
14. My major sources for this account are Izzard and Gwynne-Vaughan. For women's war autobiographies, see Tylee, pp. 184–23.
15. Wells, *Ann Veronica*, pp. 159–60.
16. Presumably Evelyn Welsford is the E. J. Welsford named in 1912 as Blackman's Research Assistant: article in the *Annals of Botany*, July 1912.
17. Quoted Izzard, p. 131.

18. Gwynne-Vaughan, p. 25.
19. Gwynne-Vaughan, p. 40.
20. Gertrude George, quoted Izzard, p. 204.
21. Gwynne-Vaughan, p. 52.
22. Izzard, p. 233.

Chapter 13

1. Carden-Coyne, p. 4. For medical women's accounts, see Powell.
2. Willoughby Hyett Dickinson, quoted Kent, *Making Peace*, pp. 87–8.
3. My main secondary sources are: Ross McKibbin's introduction to Stopes, *Married Love*; Rose, June; Green.
4. http://www.ww1.manchester.ac.uk/the-university-and-the-war/impact-of-war/, accessed 2 Dec. 2015.
5. Maude, pp. 62–4.
6. Dyhouse, 'Driving Ambitions'. For women's war autobiographies, see Tylee, pp. 184–23.
7. Stopes, 'Inner Life', pp. 205–9 (quotation p. 206); Hall, Ruth, pp. 110–27.
8. Carden-Coyne, pp. 222–32 (quotation p. 225).
9. Quoted in Green, p. 180 (letter to Walter Blackie of 29 Jan. 1919).
10. Geppert (quotation p. 391). Hall, *Hidden Anxieties*.
11. Downing, p. 217.
12. Daggett, pp. 96–102; Tomes. See also Powell, pp. 199–206.This is a different Lady Paget from Lady Muriel, who was mainly based in Russia.
13. Gleichen. I am extremely grateful to Lord Malcolm Davidson for contacting me about Helena Gleichen and for providing me with private documents.
14. http://www.nationalarchives.gov.uk/womeninuniform/almeric_paget_intro.htm, accessed 10 Dec. 2015. Morley, p. 220.
15. Stopes, *Married Love*, pp. 5–7.
16. Raitt (quotation p. 71); Bennett, pp. 144–59.
17. Izzard, pp. 86, 128–9, 240.
18. Rayner-Canham and Rayner-Canham, *Devotion to Their Science*, pp. 157–8.
19. IWM EMP. 64/18.
20. Morrissey; Perrone. Chick, Hume, and MacFarlane, p. 126.
21. Phillips, Henry.
22. http://www.iwm.org.uk/collections/item/object/80010018, accessed 2 Dec. 2015, recorded in 1988. Guillebaud, p. 48; Hutton, *Memories of a Doctor*, pp. 46–7. For a film on the hospital, see http://www.cam.ac.uk/research/news/from-the-front-to-the-backs-story-of-the-first-eastern-hospital.
23. Walter Rivington, quoted in MacLeod and Moseley, p. 330.
24. Oxford and Asquith, p. 79 (Elizabeth Sloan Chesser).
25. Sir James Purves-Stewart, *The Times*, 22 Mar. 1928; Dyhouse, 'Women Students', pp. 110–13.

26. Quoted in Dyhouse, 'Driving Ambitions', p. 323.

27. Stobart, p. 95. See also Powell, pp. 181–99.

28. Quoted in Geddes, 'Deeds and Words', p. 89 (from Sharp's *Unfinished Adventure*).

29. Elston. For Scharlieb, see Jones, Greta. Among many other sources, Bennett (pp. 10–54) gives a 1915 perspective, and Crawford, *Enterprising Women*, provides a detailed account of the Garretts and their circle, and the early position of women in medicine.

30. Crawford, *Enterprising Women*, pp. 49, 104–5.

31. Morley, p. 170.

32. Hutton, *Memories of a Doctor*, pp. 1–107.

33. Daggett, p. 74.

34. Dyhouse, *No Distinction of Sex*, pp. 26–7.

35. Morley, pp. 137–67.

36. McLaren, Barbara, p. 4.

37. Mann, p. 62. Dyhouse, 'Driving Ambitions' and 'Women Students', Leneman, 'Medical Women', and Whitehead, pp. 107–25 are my main sources for this discussion.

38. Daggett, p. 73.

39. Quoted in Geddes, 'Deeds and Words', p. 97.

40. Stobart, p. 101.

41. Hutton, *With a Woman's Unit*, p. 26. NCC-WW, p. 3.

42. Letter of 7 Jul. 1918 from Edith Stoney (in Royaumont) to Sir James Mackenzie Davidson. I am extremely grateful to his grandson, Lord Davidson, for sending me a copy of this letter.

43. Jessie Laurie quoted in Leneman, *In the Service of Life*, p. 30.

44. McLaren, pp. 41–6; Leneman, 'Medical Women', pp. 163–4. Mann, pp. 29–31.

45. For other women's hospitals, see Bennett, pp. 57–159.

46. McLaren, Barbara, pp. 1–4, Murray, Flora and Geddes 'Deeds and Words' are my main sources for the Endell Street Hospital.

47. White, pp. 139–44.

48. Murray, Flora, pp. 172–3. See also Powell, pp. 68–72.

49. Geddes, 'Doctor's Dilemma'. See also Whitehead, pp. 104–6.

50. Quoted Geddes, 'Deeds and Words', p. 96 (1916).

51. Pankhurst, Sylvia, pp. 54–5.

52. Quoted in Carden-Coyne, p. 197 (Middlesex Hospital).

53. Quoted in Geddes, 'Deeds and Words', p. 85 (Marion Dickerman).

54. Sinclair, pp. 27, 47 (on a Belgian military hospital).

55. Murray, Flora, p. 129.

56. Mohr. Dr W. F. Buckley from Newnham worked there as a surgeon: NCC-WW, p. 1.

57. Bennett, pp. 58–9.

58. Quoted in Geddes, 'Deeds and Words', p. 88.

59. Carden-Coyne, pp. 191–337.
60. Geddes, 'Doctors' Dilemma', pp. 213–14.
61. Michaelson (quotation p. 171).

Chapter 14

1. Leneman, *Service of Life*. Powell, pp. 169–81.
2. Hutton, *With a Woman's Unit*, p. 160.
3. Elsie Bowerman quoted in Kent, *Making Peace*, p. 59 (diary entry, 1 Jan. 1917).
4. Näpflin.
5. LSE: 7/BSH/2/2/5, letters from Ellie Rendel to Ray Strachey of 8 and 24 Sep. 1916.
6. http://www.scotsman.com/lifestyle/dr-elizabeth-ross-the-scottish-saint-of-serbia-1-789188, accessed 5 Jan. 2016.
7. Leneman, 'Medical Women', pp. 168–70.
8. Matthews.
9. Whitehead, pp. 139–51. Powell, pp. 365–8.
10. Conway, pp. 1054–64; Whitehead, pp. 107–25 (Dr Norman Walker quoted on p. 107). Strangely, Mark Harrison does not mention women's initiatives.
11. Isabel Emslie, quoted in Leneman, 'Medical Women', p. 172.
12. LSE: 7/BSH/2/2/5, letter from Ellie Rendel to Ray Strachey, 8 Sep. 1916.
13. LSE: 7/BSH/2/2/5, letter of 26 Mar. 1917 from Ellie Rendel to Ray Strachey (from Jasi).
14. Hutton, *A Woman's Unit*, dedication.
15. My main biographical sources are: Hutton, *A Woman's Unit* and *Memories of a Doctor*; Hutton's obituary, *The Times*, 12 Jan. 1960; McConnell. For women's war autobiographies, see Tylee, pp. 184–23.
16. Emslie.
17. Harrison, Mark, pp. 228–39 (John Dillon quoted on p. 235). See also http://www.1914-1918.net/salonika.htm, accessed 11 Jan. 2016.
18. LSE: 7/BSH/2/2/5, letter of 29 Oct. 1916 from Ellie Rendel to Ray Strachey (from 1st Serbian Field Hospital).
19. Quoted in Leneman, *In the Service of Life*, p. 159.
20. Quoted in Leneman, *In the Service of Life*, p. 175.
21. Hutton, *Memories of a Doctor*, p. 166.
22. Letter from May Green to Jessie Laurie quoted in Leneman, *In the Service of Life*, p. 190.
23. http://www.medicinskavranje.edu.rs/lat/%C5%A1kola, accessed 13 Jan. 2016. I am grateful to Dr Sarah Hutton for this web reference.
24. Quoted in Leneman, *In the Service of Life*, p. 175.
25. Quoted in Leneman, *In the Service of Life*, pp. 198, 199.
26. Dr Ruth Conway quoted in Leneman, *In the Service of Life*, pp. 197–8.
27. Quoted in Leneman, *In the Service of Life*, p. 200.

28. Hutton, *Memories of a Doctor*, p. 186.
29. This unit was established by Lady Muriel Paget (Lady Leila Paget was in Skopje). Powell, pp. 68–72.
30. Hutton, *Memories of a Doctor*, pp. 212–13.
31. http://www.ucl.ac.uk/bloomsbury-project/institutions/british_brain_hospital. htm, accessed 14 Jan. 2016. Hutton, *Memories of a Doctor*, pp. 255–64.
32. Correspondence in *The Times* of 22, 23, and 30 Mar. 1928.
33. Geddes, 'Doctors' Dilemma', pp. 207–8.
34. Hutton, *Memories of a Doctor*, p. 217. Hall, *Hidden Anxieties*.
35. Hutton, *Memories of a Doctor* (quotation p. 248) and *Hygiene of Marriage*; Connell and Hunt; Stopes, *Married Love*, including Ross McKibbin's introduction.
36. Rose West quoted in Leneman, *In the Service of Life*, p. 175.
37. Hutton, *Memories of a Doctor*, p. 224.
38. Gandhi quoted in Hutton, *Memories of a Doctor*, pp. 300–1.

Chapter 15

1. Gottlieb and Toye (quotation p. 7). For a recent analysis of the extensive literature, see Pugh, pp. 32–218.
2. Jones, Allan (Reith quoted on p. 971). I am very grateful to Robert Bud of the Science Museum for alerting me to Mary Adams. Her contemporary Hilda Matheson, an Oxford history graduate who had worked for MI5 during the War, also helped to develop the concept of the scripted talk.
3. Richard Lambert quoted in Jones, Allan, p. 975.
4. Jones, Allan (quotation p. 974).
5. Noakes, '"Playing at Being Soldiers"' (quotation p. 143 from *The Times*, 28 Aug. 1919).
6. Quoted in Thom, *Nice Girls*, p. 41 (c.1919).
7. Quoted in Pursell, p. 83 (3 June 1919).
8. Clarsen (1982 quotation p. 348).
9. Quoted Braybon, *Women Workers*, p. 185 (from *The Engineer*, 14 Nov. 1919).
10. Kathleen Culhane quoted in Horrocks, p. 359.
11. Quoted in Nicholson, p. 159 (*Westminster Gazette*).
12. Hall, 'Chloe, Olivia', pp. 198–9 (quotation p. 199 from a letter to Edward Mellanby).
13. Cicely Hamilton quoted in Whitelaw, p. 64 (from *Life Errant*). Gotha was an important industrial centre in Germany.
14. Cicely Hamilton quoted in Whitelaw, p. 62 (from *Life Errant*).
15. Dyhouse, *Girl Trouble*, p. 85 (on Winifred Foley).
16. Quoted in Braybon, *Women Workers*, p. 166.
17. Quoted in Grayzel, p. 135 (from *The Woman Worker*, Jun. 1918).

18. Caine, *English Feminism*, pp. 173–221. Hamilton quoted from *Life Errant* in Rayner-Canham and Rayner-Canham, *Chemistry was Their Life*, pp. 474–5.
19. Conway, p. 1058.
20. Noakes, '"Playing at Being Soldiers"' (*Saturday Review* of 11 Jan. 1919 quoted p. 143).
21. Noakes, *Women in the British Army*, p. 85 (Irene Clapham, 1935). Dyhouse, *Girl Trouble*, pp. 70–104.
22. Quoted in Nicholson, p. 31 (original emphasis). For the civil service, see Zimmeck.
23. Dame Lilian Baker quoted in Adlington, p. 27.
24. IWM EMP. 64/17.
25. IWM EN1/3/COR/5 (letter from Miss S. Wolfe-Murray to Agnes Conway of 19 Oct. 1917, emphasis in original).
26. Sir George Barnes quoted in Adams, p. 112 (*The Times*, 12 Aug. 1920).
27. *The Times*, 29 Nov. 1919, p. 13.
28. Gleichen, p. 310.
29. IWM EMP. 64/39.
30. McLaren, Angus; Connell and Hunt; Hall, *Hidden Anxieties*.
31. Kent, *Making Peace*, pp. 97–143 (Kenealy quoted on p. 113).
32. Margaret Partridge to Caroline Haslett in 1925, quoted in Pursell, p. 93.
33. E. Neville (1933), quoted in Rayner-Canham and Rayner-Canham, *Chemistry was Their Life*, p. 479.
34. Nicholson, p. 328.
35. Pick, pp. 231–4; Showalter, pp. 67–94.
36. McLaren, Angus.
37. Edgerton, *Warfare State*. MacLeod, Christine.
38. IWM ED. 2. 16/928 Aug. 1919, letter from Fred Knowles, Sheffield University metallurgy department.
39. Forster (quotation p. 11).
40. Horrocks (careers journal of 1938 quoted p. 352).
41. Jordan Lloyd quoted in Rayner-Canham and Rayner-Canham, *Chemistry was Their Life*, p. 477.
42. Horrocks (Kathleen Culhane quoted on p. 359).
43. Richard Pilcher quoted in Rayner-Canham and Rayner-Canham, *Chemistry was Their Life*, p. 476.
44. Dyhouse, 'Women Students'. Whitehead, pp. 107–25.
45. Hutton, *Memories of a Doctor*, p. 177.
46. Ina Brooksbank quoted in Rayner-Canham and Rayner-Canham, *Chemistry was Their Life*, p. 472.
47. Wang, especially p. 123.
48. Dyhouse, *No Distinction of Sex*, pp. 238–51 (quotation p. 242).

Chapter 16

1. https://royalsociety.org/news/2010/influential-british-women/, accessed 5 June 2014.
2. Burstall, p. 2946.
3. Rayner-Canham and Rayner-Canham, *Chemistry was Their Life*, pp. 485–8.
4. Quoted in Ogilvie and Harvey, vol. 1, p. 69.
5. Quoted in Fara, p. 165.
6. Caroline Herschel's earlier publication was only a short note on observing a comet.
7. Quoted in Fara, pp. 101–2.

BIBLIOGRAPHY

Adam, Ruth. *A Woman's Place: 1910–1975* (London: Persephone Books, 2000).

Adie, Kate. *Fighting on the Home Front: The Legacy of Women in World War One* (London: Hodder & Stoughton, 2013).

Adlington, Lucy. *Fashion: Women in World War One* (Andover: Pitkin Publishing, 2014).

Alec-Tweedie, Ethel. *Women and Soldiers* (London: John Lane, the Bodley Head, 1918).

Baker, Nina. 'Early Women Engineering Graduates from Scottish Universities', *Women's History Magazine* 60 (2009), pp. 21–30.

Barrett, Anne. 'Whiteley, Martha Annie', *Oxford Dictionary of National Biography*, accessed 9 July 2013.

Bennett, Alice Horlock. *English Medical Women: Glimpses of Their Work in Peace and War* (London: Pitman, 1915).

Bidwell, Shelford. *The Women's Royal Army Corps* (London: Cooper, 1977).

Bindman, Lynn, Alison Brading, and Tilli Tansey (eds). *Women Physiologists: An Anniversary Celebration of Their Contributions to British Physiology* (London and Chapel Hill: Portland Press, 1993).

Birch, Frank, Dilly Knox, and G. P. Mackeson. *Alice in I.D.25: A Codebreaking Parody of Alice's Adventures in Wonderland* (Stroud: Pitkin Publishing, 2015).

Bland, Lucy. *Banishing the Beast: Sexuality and the Early Feminists* (New York: New Press, 1995).

Bogen, Anna. *Women's University Fiction, 1880–1945* (London: Pickering & Chatto, 2014).

Bostridge, Mark. *The Fateful Year: England 1914* (London: Penguin 2014).

Bowden, Ruth E. M. 'Cullis, Winifred Clara', *Oxford Dictionary of National Biography*, accessed 4 August 2015.

Braybon, Gail. *Women Workers in the First World War: The British Experience* (London: Croom Helm, 1981).

Braybon, Gail. 'Women, War, and Work', in Hew Strachan (ed.), *The Oxford Illustrated History of the First World War* (Oxford: Oxford University Press, 1998), pp. 148–62.

Braybon, Gail. 'Winners or Losers: Women's Symbolic Role in the War Story', in Gail Braybon (ed.), *Evidence, History and the Great War: Historians and the Impact of 1914–18* (New York and Oxford: Berghahn Books, 2003), pp. 86–112.

Braybon, Gail and Penny Summerfield. *Out of the Cage: Women's Experiences in Two World Wars* (London: Pandora, 1987).

Brittain, Vera. *Testament of Youth* (London: Fontana, 1979).

Brockington, Grace. 'Smedley, (Annie) Constance', *Oxford Dictionary of National Biography*, accessed 4 August 2015.

Browne, Janet. 'Pertz, Dorothea Frances Matilda', *Oxford Dictionary of National Biography*, accessed 28 July 2015.

Burek, C. V. 'The Role of Women in Geological Higher Education—in Bedford College, London (Catherine Raisin) and Newnham College, Cambridge, UK', in C. V. Burek and B. Higgs (eds), *The Role of Women in the History of Geology* (London: The Geological Society, 2007), pp. 9–38.

Burstall, F. H. 'Frances Mary Gore Micklethwait (1868–1950)', *Journal of the Chemical Society* (1952), pp. 2946–7.

Caine, Barbara. *English Feminism, 1780–1980* (Oxford: Oxford University Press, 1997).

Caine, Barbara. *Bombay to Bloomsbury: A Biography of the Strachey Family* (Oxford: Oxford University Press, 2005).

Carden-Coyne, Ana. *The Politics of Wounds: Military Patients and Medical Power in the First World War* (Oxford: Oxford University Press, 2014).

Cardwell, D. S. A. 'A Discussion on the Effects of the Two World Wars on the Organization and Development of Science in the United Kingdom', *Proceedings of the Royal Society of London. Series A, Mathematical and Physical Sciences* 342 (1975), pp. 447–56.

Chadwick, Whitney. 'The Fine art of Gentling: Horses, Women and Rosa Bonheur in Victorian England', in Kathleen Adler and Marcia Pointon (eds), *The Body Imaged: The Human Form and Visual Culture since the Renaissance* (Cambridge: Cambridge University Press, 1993), pp. 89–107.

Chick, Harriette, Margaret Hume, and Marjorie MacFarlane. *War on Disease: A History of the Lister Institute* (London: Andre Deutsch, 1971).

Clark, J. F. M. 'Ormerod, Eleanor', *Oxford Dictionary of National Biography*, accessed 18 July 2015.

Clarsen, Georgine. '"A Fine University for Women Engineers": A Scottish munitions Factory in World War I', *Women's History Review* 12 (2003), pp. 333–56.

Cocroft, Wayne D. 'First World War Explosives Manufacture: The British Experience', in Roy MacLeod and Jeffrey Allan Johnson (eds), *Frontline and Factory: Comparative Perspectives on the Chemical Industry at War, 1914–1924* (Dordrecht: Springer, 2006), pp. 31–46.

Cohen, Susan. *Medical Services in the First World War* (Oxford: Shire Publications, 2014).

Condell, Diana and Jean Liddiard. *Working for Victory? Images of Women in the First World War, 1914–18* (London: Routledge & Kegan Paul, 1987).

Connell, Erin and Alan Hunt. 'Sexual Ideology and Sexual Physiology in the Discourses of Sex Advice Literature', *Canadian Journal of Human Sexuality* 15 (2006), 23–45.

Conway, Agnes. 'Women's War-Work', *Encyclopaedia Britannica* (1922), vol. 23, pp. 1054–64.

Crawford, Elizabeth. *Enterprising Women: The Garretts and Their Circle* (London: Francis Boutle, 2002).

Crawford, Elizabeth (ed.). *Campaigning for the Vote: Kate Parry Frye's Suffrage Diary* (London: Francis Boutle, 2013).

Creese, Mary R. S. 'British Women of the Early Nineteenth and Early Twentieth Centuries who Contributed to Research in the Chemical Sciences', *British Journal for the History of Science* 24 (1994), 275–305.

Creese, Mary R. S. 'Whiteley, Martha Annie, Chemist and Editor', *Bulletin of the History of Chemistry* 20 (1997), pp. 42–5.

Creese, Mary R. S. 'Benson, Margaret', *Oxford Dictionary of National Biography*, accessed 24 August 2015.

Creese, Mary R. S. 'Hudson, Hilda', *Oxford Dictionary of National Biography*, accessed 27 July 2015.

Creese, Mary R. S. 'Maclean, Ida Smedley', *Oxford Dictionary of National Biography*, accessed 3 August 2015.

Creese, Mary R. S. 'Sargant, Ethel', *Oxford Dictionary of National Biography*, accessed 28 July 2015.

Creese, Mary R. S. 'Shakespear, Ethel Mary Reader', *Oxford Dictionary of National Biography*, accessed 7 August 2015.

Creese, Mary R. S. 'Smith, Annie Lorrain', *Oxford Dictionary of National Biography*, accessed 28 July 2015.

Creese, Mary R. S. 'Thomas, Ethel', *Oxford Dictionary of National Biography*, accessed 7 August 2015.

Daggett, Mabel Potter. *Women Wanted: The Story Written in Blood Red Letters on the Horizon of the Great World War* (New York: George H. Doran, 1918).

Davis, A. E. L. 'Cave-Browne, Beatrice Mabel', *Oxford Dictionary of National Biography*, accessed 27 July 2015.

Deacon, Margaret. 'Marshall, Sheina Macalister', *Oxford Dictionary of National Biography*, accessed 27 July 2015.

de Havilland, Gladys. 'The "Fannys"', in Cardinal, Agnès, Dorothy Goldman, and Judith Hattaway (eds), *Women's Writing on the First World War* (Oxford: Oxford University Press, 1999), pp. 192–7 (from *The Woman's Motor Manual*, 1915).

de Vries, Jacqueline. 'Gendering Patriotism: Emmeline and Christabel Pankhurst and World War One', in Oldfield, Sybil (ed.), *This Working-Day World: Women's Lives and Culture(s) in Britain 1914–1945* (London: Taylor & Francis, 1994), pp. 75–88.

Dillon, Brian. *The Great Explosion: Gunpowder, the Great War, and a Disaster on the Kent Marshes* (London: Penguin Books, 2015).

Doran, Amanda-Jane and Andrew McCarthy. *The Huns have got my Gramophone! Advertisements from the Great War* (Oxford: Bodleian Library, 2014).

Downing, Taylor. *Secret Warriors: Key Scientists, Code-Breakers and Propagandists of the Great War* (London: Little, Brown, 2014).

Drake, Barbara. *Women in the Engineering Trades: A problem, a Solution, and Some Criticisms* (London: Fabian Research Department and George Allen & Unwin, 1917).

Drake, Barbara. *Women in Trade Unions* (Labour Research Department and George Allen & Unwin, 1921).

Drummond, Cherry. *The Remarkable Life of Victoria Drummond, Marine Engineer* (London: Institute of Marine Engineers, 1994).

Dyhouse, Carol. *No Distinction of Sex? Women in British Universities 1870–1939* (London: UCL Press, 1995).

Dyhouse, Carol. 'The British Federation of University Women and the Status of Women in Universities, 1907–1939', *Women's History Review* 4 (1995), pp. 465–85.

Dyhouse, Carol. 'Driving Ambitions: Women in Pursuit of a Medical Education, 1890–1939', *Women's History Review* 7 (1998), pp. 321–43.

Dyhouse, Carol. 'Women Students and the London Medical Schools, 1914–39: The Anatomy of a Masculine Culture', *Gender & History* 10 (1998), pp. 110–32.

Dyhouse, Carol. *Students: A Gendered History* (London: Routledge, 2006).

Dyhouse, Carol. *Girl Trouble: Panic and Progress in the History of Young Women* (London and New York: Zed Books, 2013).

Edgerton, David. *England and the Aeroplane: An Essay on a Militant and Technological Nation* (Basingstoke: Macmillan, in association with the Centre for the History of Science, Technology and Medicine, University of Manchester, 1991).

Edgerton, David. *Science, Technology, and the British Industrial 'Decline', 1870–1970* (Cambridge: Cambridge University Press for the Economic History Society, 1996).

Edgerton, David. *The Shock of the Old: Technology and Global History since 1900* (London: Profile, 2006).

Edgerton, David. *Warfare State: Britain, 1920–1970* (Cambridge: Cambridge University Press, 2006).

Edgerton, David. 'Time, Money, and History', *Isis* 103 (2012), pp. 316–27.

Edgerton, David and John V. Pickstone. 'Science, Technology and Medicine in the United Kingdom, 1750–2000', in Ron Numbers (ed.), *The Cambridge History of Science, Volume 8: Modern Science in National and International Context* (Cambridge: Cambridge University Press, forthcoming).

Elston, Mary Ann. '"Run by Women, (mainly) for Women": Medical Women's Hospitals in Britain, 1866–1948', in Lawrence Conrad and Anne Hardy (eds), *Women and Modern Medicine* (Amsterdam: Rodopi, 2001), pp. 73–107.

Emling, Shelley. *Marie Curie and Her Daughters: The Private Lives of Science's First Family* (New York: Palgrave Macmillan, 2012).

Emslie, Isabel. 'The War and Psychiatry', *Edinburgh Medical Journal* 14 (1915), pp. 359–67.

Evans, Joan. *The Conways: A History of Three Generations* (London: Museum Press, 1966).

Fara, Patricia. *Pandora's Breeches: Women, Science and Power in the Enlightenment* (London: Pimlico, 2004).

Fawcett, Jane. 'McKenny Hughes, Thomas', *Oxford Dictionary of National Biography*, accessed 19 July 2016.

Ford, Anna Leendertz. 'Tipper, Constance Fligg', *Oxford Dictionary of National Biography*, accessed 28 July 2015.

Forster, Emily L. B. *Analytical Chemistry as a Profession for Women* (London: Charles Griffen, 1920).

Fraser, William Lionel. *All to the Good* (London: Heinemann, 1963).

Freedman, Robert. 'Ida Smedley MacLean (1877–1944): Pioneering Biochemist and Feminist Campaigner', unpublished paper presented at the ESHS conference in Prague, September 2016.

Freemantle, Michael. *Gas! Gas! Quick Boys: How Chemistry Changed the First World War* (Stroud: Spellmount, 2012).

Gay, Hannah. *The History of Imperial College London, 1907–2007: Higher Education and Research in Science, Technology and Medicine* (London: Imperial College Press, 2007).

Geddes, Auckland. *The Forging of a Family: A Family Story Studied in its Genetical, Cultural, and Spiritual Aspects, and a Testament of Personal Belief Founded Thereon* (London: Faber and Faber, 1952).

Geddes, Jennian F. 'Deeds and Words in the Suffrage Military Hospital in Endell Street', *Medical History* 51 (2007), pp. 79–98.

Geddes, Jennian F. 'The Doctors' Dilemma: Medical Women and the British Suffrage Movement', *Women's History Review* 18 (2009), pp. 203–18.

Geppert, Alexander C. T. 'Divine Sex, Happy Marriage, Regenerated Nation: Marie Stopes's Marital Manual *Married Love* and the Making of a Best-Seller, 1918–1955', *Journal of the History of Sexuality* 8 (1998), pp. 389–433.

Gibert, Julie S. 'Women Students and Student Life at England's Civic Universities before the First World War', *History of Education* 23 (1994), pp. 405–22.

Gleichen, Helena. *Contacts and Contrasts* (London: John Murray, 1940).

Gollin, Alfred. 'The Mystery of Lord Haldane and Early British Military Aviation', *Albion: A Quarterly Journal Concerned with British Studies* 11 (1979), pp. 46–65.

Goodman, Martin. 'The High Altitude Research of Mabel Purefoy Fitzgerald, 1911–13', *Notes and Records of the Royal Society* 69 (2015), pp. 85–99.

Gottlieb, Julie V. and Richard Toye. 'Introduction', in Julie V. Gottlieb and Richard Toye (eds), *The Aftermath of Suffrage: Women, Gender, and Politics in Britain, 1918–1945* (London: Palgrave Macmillan, 2013), pp. 1–8.

Gould, Paula. 'Women and the Culture of University Physics in Late Nineteenth-Century Cambridge', *British Journal for the History of Science* 30 (1997), pp. 127–49.

Grattan-Guinness, Ivor. 'A Mathematical Union: William Henry and Grace Chisholm Young', *Annals of Science* 29 (1972), pp. 195–85.

Grayzel, Susan R. *Women and the First World War* (Harlow: Longman, 2002).

Green, Stephanie. *The Public Lives of Charlotte and Marie Stopes* (London: Pickering & Chatto, 2013).

Gregory, Adrian. 'British "War Enthusiasm" in 1914: A Reassessment', in Gail Braybon (ed.), *Evidence, History and the Great War: Historians and the Impact of 1914–18* (New York and Oxford: Berghahn Books, 2003), pp. 67–85.

Guillebaud, Philomena. *From Bats to Beds to Books: The First Eastern General Hospital (Territorial Force) Cambridge—and What Came Before and After It* (Haddenham, Cambridgeshire: Fern House, 2012).

Gwynne-Vaughan, Helen. *Service with the Army* (London: Hutchinson, 1942).

Hall, Lesley A. *Hidden Anxieties: Male Sexuality, 1900–1950* (Cambridge: Polity, 1991).

Hall, Lesley A. 'Chloe, Olivia, Isabel, Letitia, Harriette, Honor and Many More: Women in Medicine and Biomedical Science, 1914–1945', in Sybil Oldfield (ed.), *This Working-Day World: Women's Lives and Culture(s) in Britain 1914–1945* (London: Taylor & Francis, 1994), pp. 192–202.

Hall, Ruth. *Marie Stopes: A Biography* (London: Deutsch, 1977).

Hamilton, Cicely. *Marriage as a Trade* (London: Women's Press, 1981/1909).

Hamilton, Mary Agnes. *Remembering My Good Friends* (London: Cape, 1944).

Harrison, Brian. *Prudent Revolutionaries: Portraits of British Feminists between the Wars* (Oxford: Clarendon Press, 1987).

Harrison, Mark. *The Medical War: British Military Medicine in the First World War* (Oxford: Oxford University Press, 2010).

Hartcup, Guy. *The War of Invention: Scientific Developments, 1914–1918* (London: Brassey's Defence, 1988).

Hay, Alice Ivy. *Valiant for Truth: Malcolm Hay of Seaton* (London: Spearman, 1971).

Hemmungs Wirtén, Eva. *Making Marie Curie: Intellectual Property and Celebrity Culture in an Age of Information* (Chicago and London: Chicago University Press, 2015).

Holton, Sandra Stanley. *Feminism and Democracy: Women's Suffrage and Reform Politics in Britain 1900–1918* (Cambridge: Cambridge University Press, 1986).

Horrocks, Sally. 'A Promising Pioneer Profession? Women in Industrial Chemistry in Inter-War Britain', *British Journal for the History of Science* 33 (2000), pp. 351–67.

Howson, Geoffrey. 'Williams, Elizabeth', *Oxford Dictionary of National Biography*, accessed 28 July 2015.

Hull, Andrew. 'War of Words: The Public Science of the British Scientific Community and the Origins of the Department of Scientific and Industrial Research, 1914–16', *British Journal for the History of Science* 32 (1999), pp. 461–81.

Hutton, Isabel Emslie. *The Hygiene of Marriage* (London: Heinemann, 1923).

Hutton, Isabel Emslie. *With a Woman's Unit in Serbia, Salonika and Sebastopol* (London: Williams, 1928).

Hutton, Isabel Emslie. *Memories of a Doctor in War and Peace* (London: Heinemann, 1960).

Hynes, Samuel. *A War Imagined: The First World War and English Culture* (London: Bodley Head, 1990).

Izzard, Molly. *A Heroine in Her Time: A life of Dame Helen Gwynne-Vaughan, 1879–1967* (London: Macmillan, 1969).

Jaffé, Deborah. *Ingenious Women: From Tincture of Saffron to Flying Machines* (Stroud: Sutton, 2003).

James, William Milbourne. *The Eyes of the Navy: A Biographical Study of Admiral Sir Reginald Hall, K.C.M.G., C.B., LL.D., D.C.L.* (London: Methuen, 1955).

Johnson, Jeffrey Allan. 'Women in the Chemical Industry in the First Half of the 20th Century', in Renate Tobies and Annette B. Vogt (eds), *Women in Industrial Research* (Stuttgart: Franz Steiner Verlag, 2014), pp. 119–44.

Jones, Allan. 'Mary Adams and the Producer's Role in early BBC Science Broadcasts', *Public Understanding of Science* 21 (2012), pp. 968–83.

Jones, Claire G. 'The Laboratory: A Suitable Place for a Woman? Gender and Laboratory Culture around 1900', in Krista Cowman and Louise A. Jackson (eds), *Women and Work Culture: Britain c.1850–1950* (Aldershot: Ashgate, 2005), pp. 177–94.

Jones, Claire G. *Femininity, Mathematics and Science, 1880–1914* (London: Macmillan Palgrave, 2009).

Jones, Greta. 'Scharlieb, Mary', *Oxford Dictionary of National Biography*, accessed 7 December 2015.

Katzir, Shaul. 'Who Knew Piezoelectricity? Rutherford and Langevin on Submarine Detection and the invention of Sonar', *Notes and Records of the Royal Society* 66 (2012), pp. 141–57.

Kent, Susan Kingsley. *Sex and Suffrage in Britain, 1860–1914* (Princeton, NJ: Princeton University Press, 1987).

Kent, Susan Kingsley. *Making Peace: The Reconstruction of Gender in Interwar Britain* (Princeton, NJ: Princeton University Press, 1993).

Kent, Susan Kingsley. *Gender and Power in Britain, 1640–1990* (London: Routledge, 1999).

Lee, Hermione. *Virginia Woolf* (London: Chatto & Windus, 1996).

Leneman, Leah. 'Medical Women at War, 1914–1918', *Medical History* 28 (1994), pp. 160–77.

Leneman, Leah. *In the Service of Life: The Story of Elsie Inglis and the Scottish Women's Hospitals* (Edinburgh: Mercat Press, 1994).

Liddington, Jill. *Vanishing for the Vote: Suffrage, Citizenship and the Battle for the Census* (Manchester: Manchester University Press, 2014).

MacLeod, Christine. 'Reluctant Entrepreneurs: Patents and State Patronage in New Technosciences, circa 1870–1930', *Isis* 103 (2012), pp. 328–39.

MacLeod, Roy. 'The Chemists go to War: The Mobilization of Civilian Chemists and the British War Effort, 1914–1918', *Annals of Science* 50 (1993), pp. 455–81.

MacLeod, Roy and Russell Moseley. 'Fathers and Daughters: Reflections on Women, Science and Victorian Cambridge', *History of Education* 8 (1979), pp. 321–43.

Mann, Ida. *Ida and the Eye: A Woman in British Ophthalmology* (Tunbridge Wells: Parapress, 1996).

Marwick, Arthur. *The Deluge: British Society and the First World War* (London: Bodley Head, 1965).

Mason, Joan. 'A Forty Years' War', *Chemistry in Britain* 27 (1991), pp. 233–8.

Mason, Joan. 'Hertha Ayrton: A Scientist of Spirit', in Gill Kirkup and Laurie Smith Keller (eds), *Inventing Women: Science, Technology and Gender* (Cambridge: Polity Press, 1992), pp. 168–77.

Mason, Joan. 'The Women Fellows' Jubilee', *Women: A Cultural Review* 6 (1995), pp. 220–33.

Mason, Joan. 'Bidder, Marion Greenwood', *Oxford Dictionary of National Biography*, accessed 7 August 2015.

Mathews, Vera Laughton. *Blue Tapestry* (London: Hollis & Carter, 1948).

Matthews, Caroline Twigge. *Experiences of a Woman Doctor in Serbia* (London: Mills & Boon, 1916).

Maude, Aylmer. *The Authorized Life of Marie C. Stopes* (London: Williams & Norgate, 1924).

McConnell, Anita. 'Hutton, Isabel Galloway Emslie', *Oxford Dictionary of National Biography*, accessed 6 January 2016.

McKillop, M. and McKillop, A. D. *Efficiency Methods: an Introduction to Scientific Management* (London: Routledge, 1917).

McLaren, Angus. *Twentieth-Century Sexuality: A History* (Oxford: Blackwell, 1999).

McLaren, Barbara. *Women of the War* (London and New York: Hodder & Stoughton, 1917).

Messenger, Rosalind. *The Doors of Opportunity: A Biography of Dame Caroline Haslett* (London: Femina Books, 1967).

Miall, Agnes M. *The Bachelor Girl's Guide to Everything, or, The girl on Her Own* (London: S. W. Partridge, 1916).

Michaelson, Kaarin. '"Union is Strength": The Medical Women's Federation and the Politics of Professionalism, 1917–30', in Krista Cowman and Louise A. Jackson (eds), *Women and Work Culture: Britain c.1850–1950* (Aldershot: Ashgate, 2005), pp. 161–76.

Mohr, Peter D. 'Chambers, Helen', *Oxford Dictionary of National Biography*, accessed 10 December 2015.

Morley, Edith Julia (ed.). *Women Workers in Seven Professions: A Survey of Their Economic Conditions and Prospects* (London: George Routledge & Sons, 1914).

Morrissey, Shelagh. 'Dame Harriette Chick DBE (1875–1977)', in Lynn Bindman, Alison Brading, and Tilli Tansey (eds), *Women Physiologists: An Anniversary*

Celebration of Their Contributions to British Physiology (London and Chapel Hill: Portland Press, 1993), pp. 21–9.

Murray, Flora. *Women as Army Surgeons: Being the History of the Women's Hospital Corps in Paris, Wimereux and Endell Street, September 1914–October 1919* (London: Hodder & Stoughton, 1920).

Murray, Margaret Alice. *My First Hundred Years* (London: W. Kimber, 1963).

Näpflin, Maria. 'Morphine', in Agnès Cardinal, Dorothy Goldman, and Judith Hattaway (eds), *Women's Writing on the First World War* (Oxford: Oxford University Press, 1999), pp. 202–3.

Nicholson, Virginia. *Singled Out: How Two Million Women Survived Without Men after World War One* (London: Penguin, 2007).

Noakes, Lucy. *Women in the British Army: War and the Gentle Sex, 1907–1948* (London: Routledge, 2006).

Noakes, Lucy. '"Playing at Being Soldiers"?: British Women and Military Uniform in the First World War', in Jessica Meyer (ed.), *British Popular Culture and the First World War* (Leiden & Boston: Brill, 2008), pp. 123–45.

Oakley, Hilda D. 'King's College for Women', in Fossey John Cobb Hearnshaw, *The Centenary History of King's College, London, 1828–1928* (London: Harrap, 1929), pp. 489–509.

Ogilvie, Marilyn B. and Joy D. Harvey. *The Biographical Dictionary of Women in Science* (New York and London: Routledge, 2000).

Ouditt, Sharon. *Fighting Forces, Writing Women: Identity and Ideology in the First World War* (London: Routledge, 1994).

Oxford and Asquith. Countess of. *Myself When Young: By Famous Women of Today* (London: Frederick Muller, 1938).

Pankhurst, Emmeline. *Suffragette: My Own Story* (London: Hesperus, 2014/1914).

Pankhurst, E. Sylvia. 'Christmas 1914', in Agnès Cardinal, Dorothy Goldman, and Judith Hattaway (eds), *Women's Writing on the First World War* (Oxford: Oxford University Press, 1999), pp. 48–60 (from *The Home Front*, 1987).

Peck, Will. 'Ida Smedley Maclean: Cambridge, Women and Science in the First World War'. Unpublished dissertation, University of Cambridge, 2014.

Perrone, Fernanda Helen. '(Ellen) Marion Delf Smith', *Oxford Dictionary of National Biography*, accessed 9 December 2015.

Perrot, Francis. *Reporter: Being the Records of Events and Scenes and Reflections in the London of George the Fifth* (London: Hutchinson, 1938).

Phillips, Ann (ed.). *A Newnham Anthology* (Cambridge: Cambridge University Press, 1979).

Phillips, Henry. 'Lloyd, Dorothy Jordan', *Oxford Dictionary of National Biography*, accessed 27 July 2015.

Pick, Daniel. *Faces of Degeneration: A European Disorder, c.1848–c.1918* (Cambridge: Cambridge University Press, 1989).

Potter, Jane. 'Hidden Drama by British Women: Pageants and Sketches from the Great War', in Claire M. Tylee (ed.), *Women, the First World War and the Dramatic*

Imagination: International Essays (1914–1999) (Lewiston, NY: Edwin Mellen Press, 2000), pp. 105–32.

Pottle, Mark. 'McKillop, Margaret', *Oxford Dictionary of National Biography*, accessed 27 July 2015.

Powell, Anne. *Women in the War Zone: Hospital Service in the First World War* (Stroud: The History Press, 2013).

Pugh, Martin. *Women and the Women's Movement in Britain, 1914–1999* (Basingstoke: Macmillan, 2015).

Pursell, Carroll. '"Am I a Lady or an Engineer?" The Origins of the Women's Engineering Society in Britain, 1918 –1940', *Technology and Culture* 34 (1993), pp. 78–97.

Raitt, Suzanne. 'Early British Psychoanalysis and the Medico-Psychological Clinic', *History Workshop Journal* 58 (2004), pp. 63–85.

Ralph, Sarah. *The Road to Bletchley Park* (Bletchley Park exhibition and brochure, 2015).

Rayner-Canham, Marlene and Rayner-Canham, Geoff. *Chemistry Was Their Life: Pioneer British Women Chemists, 1880–1949* (London: Imperial College Press, 2008).

Rayner-Canham, Marlene and Rayner-Canham, Geoffrey. *Devotion to Their Science: Pioneer Women of Radioactivity* (Philadelphia: Chemical Heritage Foundation and Montreal: McGill-Queens University Press, 1997).

Rayner-Canham, Marlene and Rayner-Canham, Geoffrey. 'British Women Chemists and the First World War', *Bulletin of the History of Chemistry* 23 (1999), pp. 20–7.

Reid, R. W. *Tongues of Conscience: War and the Scientist's Dilemma* (London: Constable, 1969).

Reilly, Catherine W. *Scars upon My Heart: Women's Poetry and Verse of the First World War* (London: Virago, 1981).

Richards, Evelyn. 'Redrawing the Boundaries: Darwinian Science and Victorian Women Intellectuals', in Bernard Lightman (ed.), *Victorian Science in Context* (Chicago and London: University of Chicago Press, 1987), pp. 119–42.

Richardson, Angelique. 'Kenealy, Arabella Madonna', *Oxford Dictionary of National Biography*, accessed 9 February 2016.

Richmond, Marsha. '"A Lab of One's Own": The Balfour Biological Laboratory for Women at Cambridge University, 1884–1914', *Isis* 88 (1997), pp. 422–55.

Richmond, Marsha. 'Women in the Early History of Genetics: William Bateson and the Newnham College Mendelians, 1900–1910', *Isis* 92 (2001), pp. 55–90.

Robert, Krisztina. '"All that is Best of the Modern Woman"? Representations of Female Military Auxiliaries in British Popular Culture, 1914–1919', in Jessica Meyer (ed.), *British Popular Culture and the First World War* (Leiden & Boston: Brill, 2008), pp. 97–122.

Rose, Hilary and Rose, Steven. *Science and Society* (Harmondsworth: Penguin, 1969).

Rose, June. 'The Evolution of Marie Stopes', in Robert A. Peel (ed.), *Marie Stopes, Eugenics and the English Birth Control Movement* (London: Galton Institute, 1997), pp. 13–26.

Rose, Norman. *Chaim Weizmann: A Biography* (London: Weidenfeld & Nicolson, 1986).

Shaw, George Bernard. 'Mrs Warren's Profession', in George Bernard Shaw, *Plays Unpleasant* (London: Penguin, 2000), pp. 179–286.

Shindler, Karolyn. *Discovering Dorothea: The Life of the Pioneering Fossil-Hunter Dorothea Bate* (London: HarperCollins, 2005).

Showalter, Elaine. *The Female Malady: Women, Madness and English Culture, 1830–1980* (London: Virago Press, 1987).

Simmons, Jack. *New University* (Leicester: Leicester University Press, 1958).

Sinclair, May. *A Journal of Impressions in Belgium* (London: Hutchinson, 1915).

Smedley, Constance. *Woman: A Few Shrieks! Setting Forth the Necessity of Shrieking till the Shrieks be Heard* (Letchworth: Garden City Press, 1907).

Smedley, Constance. *Crusaders: Reminiscences* (London: Duckworth, 1929).

Smith, Angela. 'All Quiet on the Woolwich Front: Literary and Cultural Constructions of Women Munitions Workers in the First World War', in Krista Cowman and Louise A. Jackson (eds), *Women and Work Culture: Britain c.1850–1950* (Aldershot: Ashgate, 2005), pp. 197–212.

Somerville, J. M. 'Dr Sophia Jex-Blake and the Edinburgh School of Medicine for Women, 1886–1898', *Journal of the Royal College of Physicians of Edinburgh* 35 (2005), pp. 261–7.

Stanley, Matthew. '"An Expedition to Heal the Wounds of War": The 1919 Eclipse and Eddington as Quaker Adventurer', *Isis* 94 (2003), pp. 57–89.

Stephenson, Marjorie. 'Muriel Wheldale Onslow, 1880–1932', *Biochemical Journal* 26 (1932), pp. 915–16.

Stevenson, Julie. 'Fishenden, Margaret', *Oxford Dictionary of National Biography*, accessed 7 August 2015.

Stobart, Mabel St Clair. 'A Woman in the Midst of War', in Agnès Cardinal, Dorothy Goldman, and Judith Hattaway (eds), *Women's Writing on the First World War* (Oxford: Oxford University Press, 1999), pp. 95–101 (original in *The Ladies Home Journal*, January 1915).

Stopes, Marie Carmichael. 'Inner Life', in Aylmer Maude, *The Authorized Life of Marie C. Stopes* (London: Williams & Norgate, 1924), pp. 185–213.

Stopes, Marie Carmichael. *Married Love* (Oxford: Oxford University Press, 2008).

Story, Neil R. and Molly Housego. *Women in the First World War* (Oxford: Shire Publications, 2010).

Strachey, Barbara. *Remarkable Relations: The Story of the Pearsall Smith Family* (London: Gollancz, 1980).

Strachey, Ray. *A Quaker Grandmother, Hannah Whitall Smith* (New York: Fleming H. Revell, 1914).

Strachey, Ray. *Marching On* (London: Jonathan Cape, 1923).

Strachey, Ray. *Women's Suffrage and Women's Service: The History of the London and National Society for Women's Service* (London: London and National Society for Women's Service, 1927).

Strachey, Ray. *The Cause: A Short History of the Woman's Movement in Great Britain* (Bath: Chivers, 1928/1974).

Strachey, Ray. *Millicent Garrett Fawcett* (London: J. Murray, 1931).

Strachey, Ray (ed.). *Our Freedom and Its Results, by Five Women* (London: Hogarth Press, 1936).

Sutherland, Gillian. *In Search of the New Woman: Middle-Class Women and Work in Britain, 1870–1914* (Cambridge: Cambridge University Press, 2015).

Terry, Roy. *Women in Khaki: The Story of the British Woman Soldier* (London: Columbus, 1988).

Terry, Roy. 'Watson, Alexandra Mary Chalmers', *Oxford Dictionary of National Biography*, accessed 18 August 2015.

Thom, Deborah. *Nice Girls and Rude Girls: Women Workers in World War I* (London: I.B. Tauris, 1998).

Thom, Deborah. 'Making Spectaculars: Museums and How we Remember Gender in Wartime', in Gail Braybon (ed.), *Evidence, History and the Great War: Historians and the Impact of 1914–18* (New York and Oxford: Berghahn Books, 2003), pp. 48–66.

Tickner, Lisa. *The Spectacle of Women: Imagery of the Suffrage Campaign 1907–14* (London: Chatto & Windus, 1987).

Tomes, Jason. 'Paget, Dame (Louise Margaret) Leila Wemyss', *Oxford Dictionary of National Biography*, accessed 5 January 2016.

Tullberg, Rita McWilliams. *Women at Cambridge* (Cambridge: Cambridge University Press, 1998).

Turner, Frank M. 'Public Science in Britain, 1880–1919', *Isis* 71 (1980), pp. 589–608.

Tylee, Claire M. *The Great War and Women's Consciousness: Images of Militarism and Womanhood in Women's Writings, 1914–64* (Basingstoke: Macmillan, 1990).

Van der Kloot, William. 'Lawrence Bragg's Role in the Development of Sound-Ranging in World War I', *Notes and Records of the Royal Society* 59 (2005), pp. 273–84.

Vicinus, Martha. *Independent Women: Work and Community for Single Women, 1850–1920* (Chicago: University of Chicago Press, 1985).

Wachtler, M. and C. V. Burek. 'Maria Matilda Ogilvie Gordon (1864–1939): A Scottish Researcher in the Alps', in C. V. Burek and B. Higgs (eds), *The Role of Women in the History of Geology* (London: The Geological Society, 2007), pp. 305–17.

Wang, Zouyue. 'The First World War, Academic Science and the "Two Cultures": Educational Reforms at the University of Cambridge', *Minerva* 33 (1995), pp. 107–27.

Wells, H. G. *Ann Veronica* (Thirsk: House of Stratus, 2002/1909).

Wells, H. G. *A Modern Utopia* (London: Penguin, 2005/1905).

West, Rebecca. 'The Cordite Makers', in Agnès Cardinal, Dorothy Goldman, and Judith Hattaway (eds), *Women's Writing on the First World War* (Oxford: Oxford University Press, 1999), pp. 91–4.

White, Jerry. *Zeppelin Nights: London in the First World War* (London: Vintage Books, 2015).

Whitehead, Ian R. *Doctors in the Great War* (London: Leo Cooper, 1999).

Whitelaw, Lis. *The Life and Rebellious Times of Cicely Hamilton* (Columbus: Ohio State University Press, 1991).

Woolf, Virginia. *A Room of One's Own* (Harmondsworth: Penguin, 1945).

Woolf, Virginia. *The Diary of Virginia Woolf, 1925–30* (ed. Anne Oliver Bell) (Harmondsworth: Penguin, 1982).

Woollacott, Angela. *On Her Their Lives Depend: Munitions Workers in the Great War* (Berkeley and London: University of California Press, 1994).

Wosk, Julie. *Women and the Machine: Representations from the Spinning Wheel to the Electronic Age* (Baltimore and London: Johns Hopkins University Press, 2001).

Wright, Almroth E. *The Unexpurgated Case Against Woman Suffrage* (New York: Paul B. Hoeber, 1913).

Wyllie, James and Michael McKinley. *Codebreakers: The Secret Intelligence Unit that Changed the Course of the First World War* (London: Ebury Press, 2015).

Wyse Jackson, Patrick N. and Mary E. Spencer Jones. 'The Quiet Workforce: The Various Roles of Women in Geological and Natural History Museums during the Early to Mid-1900s', in C. V. Burek and B. Higgs (eds), *The Role of Women in the History of Geology* (London: The Geological Society, 2007), pp. 97–113.

Zangwill, Edith Ayrton. *The Call* (London: Allen & Unwyn, 1924).

Zimmeck, Mata. 'The "New Woman" in the Machinery of Government: A Spanner in the Works?', in Roy MacLeod (ed.), *Government and Expertise: Specialists, Administrators and Professionals 1860–1919* (Cambridge: Cambridge University Press, 1988), pp. 185–202.

PICTURE CREDITS

INDEX

Note: f indicates the figure in the respective page number

Board of Invention and Research (BIR),
    Royal Navy 142
Boer Wars 41
Boole, Mary 95
Boy's Brigade 54
Bragg, William Lawrence 121, 137–8,
    145–6, 180
Bridport companies 185
Bristol University 23, 143
British Association for the Advancement of
    Science 58, 131
British Dyestuffs Corporation 272–3
British Federation of University Women
    (BFUW) 23, 132–3, 174, 176, 195
British Hospital for Mental and Nervous
    Disorders, Camden 258
British Medical Journal 77
British Westinghouse Electrical &
    Manufacturing Company 156, 274
British Women's Medical Federation 260
Brittain, Vera 12, 24, 25, 28, 81, 216–17
Brooke, Rupert 135
Bulkley, Margaret (James Barry) 226
Burt, Cyril 266

Cambridge Scientific Instrument Co 184
Cambridge University 21–2, 97–8,
    121–2, 126
    Cavendish Laboratory 121, 128, 138, 281
    see also Newnham College, Cambridge
Campbell, Janet 7
Carlyle, Thomas 10
Carroll, Lewis 189
Cat and Mouse Act 67
Cause, The (Strachey) 92
Cave-Browne, Beatrice Mabel 184
Cavell, Edith 216–17
Cavendish Laboratory, Cambridge 121, 128,
    138, 281
Cavendish, Margaret 280
censorship 183, 194
Chalmers-Watson, Alexandra Mary (Mona
    Geddes) 197, 198, 199–201, 202–4
Chalmers-Watson, Douglas 203
Chamberlain, Austen 45
Chambers, Helen 234–5
Chambers, Hettie 222
Charge of the Light Brigade, The
    (Tennyson) 137
Cheltenham Ladies' College 163, 206
chemical industry 268, 277, 278
Chemical Society 175, 178–80

chemical warfare 81–2
chemistry 157–8, 163
Chemistry and Industry (magazine) 166
Chick, Harriet 128–9, 161, 223
Chisholm, Grace 118–19, 126–7
Churchill, Winston 29, 71, 141, 147
Clarke's Business College, London 5
class differences 14, 15–16, 60, 74, 215, 270
    and eugenics 42–5, 266
    in wartime 17–19, 79–80, 276
Clayburn, Miss 158
Collins, Wilkie 30
Conan Doyle, Arthur 104
conscientious objectors (COs) 105, 106
conscription 42
Conway, Agnes 149, 150–1, 177, 236, 272–3
Conway, Martin 150
Cork Street Code Breakers 190–1
Costelloe, Frank 95–6
Costelloe, Karin, *see* Stephen, Karin (née
    Costelloe)
Costelloe, Mary 95–7
Costelloe, Rachel (Ray), *see* Strachey, Ray
    (Rachel, née Costelloe)
Council of Women Civil Servants 111
Creak, Miss (headmistress of King Edward
    VI Grammar School) 170
Crimea 245, 256–7
cryptographers/cryptography 166–7,
    185–91
Culhane, Kathleen 281–2
Cullis, Winifred 129, 174
Cummings, Miss 158
Curie, Irène 67
Curie, Marie 65, 67, 120, 176–7, 209, 222–3
    and Chemical Society 178–9
    Marie Curie Radium banner 66*f*
Curie, Pierre 210
cycling 58–9

Daggett, Mabel 78
Daily Chronicle 76
Daily Express 91
Daily Mail 61, 71, 144
Daily News 267
Darwin, Charles 32–5, 36, 117
Darwin, Erasmus 33
Darwin, Francis 118
Davidson, Gladys 78
Davison, Emily Wilding 50
Delf, Marion 223
Denniston, Alastair 188